Preface

Amphibians have a long history of use in biomedical research, but have received little formal attention as laboratory animals. The need for amphibians and the increasing difficulty encountered in obtaining suitable animals collected from the wild have produced a demand for recommendations for the breeding, care, and management of these laboratory animals. With the exception of useful but limited chapters in books on other topics and of privately circulated handout sheets, this document is a first attempt to assemble the available, practical information on the breeding, care, and management of amphibians for the laboratory. Major emphasis is on the anurans (frogs and toads) used by investigators in North America, with some attention given to the urodeles (salamanders and newts).

In many respects, this effort is premature. Though the situation is somewhat better for urodeles, so recent is the attempt to maintain anurans in long-term laboratory culture that relatively little is known about the critical factors involved. Adequate objective criteria on which to base the evaluation of quality or uniformity are simply unavailable. Although there has been an increase in the knowledge of amphibian genetics recently, the control or reproductive cycles, the evaluation of nutritional regimens, the regulation of growth, and the management of disease in amphibians are only now under serious concerted investigation.

As a consequence of these deficiencies in our knowledge, this guide is heavily dependent on the lore that has developed among amphibian biologists. Therefore, the recommendations and guidelines suggested here should be considered as tentative. Specific procedures are subject to modification in accordance with individual situations, colony size, and personal preference.

A serious handicap to the development of adequate care of amphibians is the relative inaccessibility to investigators of jargon-free, yet adequate, accounts of the "whole animal" biology of amphibians. While this document cannot contain more than brief reference to some of these matters, we hope it will provide essential information to those who are not experienced with amphibians. We also hope that it will lead those with greater experience to devise new procedures or discover new principles, whose adoption would greatly improve breeding, care, and management of amphibians for laboratory purposes.

This document should serve as a useful guide to all users of amphibians, should lead to success in the normal maintenance of amphibian colonies, and should stimulate efforts toward improving the quality of utilization of these animals. In addition to guidelines for animal care and quality, certain terminology is suggested. If accepted in standard practice, this terminology should lead to more reliable products from biomedical investigators using amphibians.

**SUBCOMMITTEE ON AMPHIBIAN STANDARDS
COMMITTEE ON STANDARDS**

GEORGE W. NACE, *Chairman*, University of Michigan
DUDLEY D. CULLEY, Louisiana State University
MARVIN B. EMMONS, E. G. Steinhilber and Company
ERICH L. GIBBS, Ultrascience Inc.
VICTOR H. HUTCHISON, University of Oklahoma
ROBERT G. McKINNELL, University of Minnesota

Contents

I **Introduction**   1
   A. General Comments   1
   B. The Need   1
      1. The Demand, 1
      2. The Supply, 3
         a. The Natural Resource, 3
         b. Artificial Culture, 3
      3. Recommendations, 7
   C. Users of Amphibians   7
   D. Laboratory-Defined Amphibians   11

II **Classification and Description of Amphibians Commonly Used for Laboratory Research**   12
   A. Phylogeny and Classification   12
   B. Description of Species Commonly Used in Laboratory Research   13
      1. Urodeles, 13
         a. *Ambystoma*, 13
         b. *Notophthalmus viridescens*, 15
         c. *Necturus maculosus*, 16
      2. Anurans, 16
         a. *Xenopus laevis*, 16
         b. *Rana catesbeiana*, 17
         c. *Rana grylio*, 18
         d. *Rana clamitans*, 18
         e. *Rana pipiens*, 21
         f. *Rana palustris*, 23
         g. *Rana sylvatica*, 24
         h. *Bufo marinus*, 24
         i. *Bombina orientalis*, 25

III Definition and Description of Experimental Amphibians 27
A. Introduction 27
B. Definition 28
1. Wild, 28
2. Wild Caught, 28
   a. Nonconditioned, 28
   b. Conditioned, 30
3. Laboratory Reared, 31
   a. Standard, 31
   b. Miscellaneous, 31
4. Laboratory Bred, 32
   a. Standard, 32
   b. Miscellaneous, 32
C. Description of Laboratory-Reared and Laboratory-Bred Amphibians 32
1. Types of Populations and Lines, 32
   a. Random Mating Lines, 33
   b. Heterozygous Isogenic Clones, 33
   c. Heterozygous Marked Lines, 33
   d. Mutant Lines, 34
   e. Inbred Lines, 34
   f. Gynogenetic Diploid Lines, 34
   g. Homozygous Lines, 34
   h. Haploid Animals, 35
   i. Polyploid Animals, 35
2. Sex Determination and Its Manipulation, 36
3. Species of Laboratory-Reared or Laboratory-Bred Amphibians Available in the United States, 37
   a. Anurans, 37
   b. Urodeles, 38

IV Sources 39
A. Selection 39
B. Dealer Care 40
C. Ordering, Shipping, and Receiving 40
1. Ordering, 40
2. Shipping, 41
3. Receiving, 42
   a. Aquatic Amphibians, 42
   b. Terrestrial Amphibians, 43
D. Legal Aspects 45

V Physical Facilities 46
A. Relation to Natural Habitat 46
B. The Amphibian Quarters 47
1. General Description, 47
   a. Isolation Quarters, 47
   b. Heating, Ventilation, and Size Specifications for Rooms, 48

   c. Description of Ancillary Rooms, 49
   d. General Specifications, 49
  2. Environmental Control, 50
   a. Water, 50
   b. Temperature, 57
   c. Lighting, 57
 C. Enclosures   58
  1. Embryos: Fertilization through Initiation of Feeding, 58
  2. Larvae, 59
  3. Juveniles, 60
   a. At Metamorphic Climax, 60
   b. Postmetamorphic, 61
  4. Adult Enclosures (Evaluation Criteria), 62
  5. Hibernation Quarters, 63
  6. Enclosure Designs, 64
   a. The Amphibian Facility of The University of Michigan, 64
   b. The *R. catesbeiana* Facility at Louisiana State University, 68
   c. Southern Frog Company, 71
   d. The Aquatic Animal Facility of Arizona State University, 72

VI Amphibian Management and Laboratory Care   75
 A. General Comments   75
 B. Anurans   76
  1. Ranidae, 76
   a. *R. pipiens,* 76
   b. *R. catesbeiana,* 84
   c. Other Ranid Species, 87
  2. Other Anurans, 88
   a. *Xenopus,* 88
   b. *Bufo,* 90
   c. *Bombina orientalis,* 90
 C. Urodela   91
  1. Axolotls, 91
   a. Early Larvae, 91
   b. Mature Larvae, 92
  2. Other Urodeles, 93

VII Breeding   95
 A. Anurans   95
  1. General Comments, 95
  2. Artificial Induction of Ovulation, 95
  3. Amplexus, 96
  4. Artificial Insemination, 97
  5. Parthenogenetic Lines and Haploid Animals, 99
  6. Homozygous Lines and Androgenesis, 100
  7. Polyploid Animals, 100
  8. Mosaic Animals, 101
  9. Nuclear Transplantation, 101
  10. *Xenopus,* 101

    B. Urodeles   103
        1. Mating, 103
        2. Spawning, 104
        3. Artificial Insemination, 106
        4. Initial Care of Embryos, 107

VIII Records and Information Control   108
    A. Identification of Individuals   108
        1. Numbering, 108
        2. Tattooing, 109
        3. Branding, 109
        4. Toe Clipping, 110
        5. Other Marking Systems, 110
        6. Drawings and Photographs, 111
    B. Information Control Systems   111

IX Amphibian Medicine   115
    A. General Comments   115
    B. Bacterial Diseases   116
        1. Pathogens, 116
        2. Drug Selection and Administration, 117
        3. Identifying Diseased Frogs, 120
        4. The Need for Treatment, 121
    C. Viral Diseases   121
        1. General Comments, 121
        2. Lucké Tumor Herpesvirus (LTHV), 122
        3. Amphibian Polyhedral Cytoplasmic Deoxyribovirus (PCDV), 122
        4. Lymphosarcoma Virus of *Xenopus,* 123
    D. Parasitic Diseases   123
    E. Mycotic Diseases   124
    F. Euthanasia and Anesthesia   124

X Personnel   127
    A. General Comments   127
    B. Safety Hazards   128
        1. Physical, 128
        2. Biological, 128

Appendix A   Status of "Endangered" Amphibians   131

Appendix B   Control Laws by States and Canadian Provinces   133

References   143

I Introduction

A. GENERAL COMMENTS

Few animals are as valuable for experimental investigations as amphibians (Kawamura and Nishioka, 1973): They are ubiquitous in nature, transitory between aquatic and terrestrial life, reasonable in size, ectothermic and the only naked vertebrates, accessible to direct examination through all developmental periods, and possess genetic mechanisms amenable to study by biological and mechanical manipulation.

Traditionally, investigators collected these animals in the wild and completed their experiments within days. In spite of their ancient history as experimental animals, care and management of these animals is rudimentary, and few established and tested guidelines for their husbandry are available (Boterenbrood, 1966; Frazer, 1966; Bardach et al., 1972). With few exceptions (e.g., mink, monkeys, hamsters, gerbils), laboratory vertebrates have been chosen from among those animals previously domesticated for agricultural purposes (e.g., chickens, goats, rabbits, swine), for pets (e.g., cats, dogs), or have been symbionts of man (e.g., mice and rats). This is not true of amphibians. While enthusiasts have maintained amphibians in limited numbers, no amphibian, except the axolotl, *Ambystoma mexicanum*, has a history of mass culture for repeated generations.

B. THE NEED

1. The Demand

Factors that stimulate a demand for amphibians are their utility for current research problems, the increased cost of avian and mammalian research animals, and the increased use of living material in high school and college instructional laboratories.

Best estimates indicate that 20 million preserved and living amphibians

TABLE 1 Common Frog (*Rana temporaria*) and Common Toad (*Bufo bufo*): A Summary of the Replies Referring to Changes in Status in Breeding Sites[a]

	1951-1955					1956-1960						1961-1965							1966-1970						
	Decrease	No Change	Increase	Total	Index of Change	Decrease	No Change	Increase	Total	Index of Change	No Longer Found	Decrease	No Change	Increase	Total	Index of Change	No Longer Found	Decrease	No Change	Increase	Total	Index of Change
Frog
Garden sites	0	8	2	10		1	8	5	14			0	6	12	12	30	+0.20	2	7	21	22	52	+0.30
Agricultural sites	5	32	0	37	−0.14	6	32	1	39	−0.13	0	22	32	4	58	−0.28	13	25	38	20	96	−0.06
Other sites	3	65	0	68	−0.04	9	68	0	77	−0.12	6	33	55	7	101	−0.27	16	57	52	19	144	−0.30
All sites	8	105	2	115	−0.05	16	108	6	130	−0.08	6	61	99	23	189	−0.21	31	39	111	61	292	−0.11
Toad
Garden sites	0	2	1	3		0	2	1	3			0	2	3	1	6		1	2	9	5	17	
Agricultural sites	2	11	0	13		2	12	0	14			1	12	8	2	23	−0.45	9	9	11	4	33	−0.21
Other sites	4	24	1	29	−0.12	6	22	1	29	−0.17	0	14	27	4	45	−0.22	6	19	24	13	62	−0.11
All sites	6	37	2	45	−0.09	8	36	2	46	−0.13	1	28	38	7	74	−0.29	16	30	44	22	112	−0.08

[a]The method of calculating the index of change is given in Cooke (1972), from which this table is adapted. Reprinted with permission of the author and the Zoological Society of London.

were used for educational purposes alone in the United States in 1971–1972. Recent studies by the Institute of Laboratory Animal Resources (ILAR) report the use of 2 million amphibians per year in research laboratories (*ILAR News*, 1972). Only rodents exceed this figure. Birds exceed it only if the number of incubated eggs are included in the tally. This heavy use of amphibians escapes the casual observer, because these animals are either used within days of their arrival in a research laboratory or lost prior to the completion of experiments due to inadequate maintenance.

2. The Supply

a. The Natural Resource

Although interest in amphibian resources is long standing (Wright, 1920), accurate figures on the status of amphibians in nature are unavailable for most American populations. This question has recently assumed great interest because of the apparent short supply and diseased state of amphibians collected from nature (Maugh, 1972; anonymous, 1973). American amphibian dealers find it increasingly difficult to identify high-density populations that permit economical collection of animals. Informal surveys among herpetologists and others interested in amphibians suggest that in recent times American amphibian populations have decreased significantly.

The best current data dealing with population reduction have been collected for British anurans (Cooke, 1972). This survey, based on questionnaires distributed to educational authorities (schools survey) and professional biologists (breeding sites survey), demonstrated that the changes in these populations were most significantly affected by habitat destruction and were directly proportional to the density of human populations and to increasing industrialization (Tables 1, 2, and 3).

Amphibian populations fluctuate drastically in response to cataclysmic environmental factors such as food shortage, drought, and early winter. In addition, the problem of short supply is aggravated by large losses that occur during shipment and holding periods (Gibbs *et al.,* 1971). However, no clear evidence exists that the survival of any of the amphibian species *commonly* used for investigative purposes is endangered by collection practices. The high reproductive potential of these amphibians assures that, given optimal environmental and climatic conditions, the populations rapidly regenerate.

b. Artificial Culture

The demand described above and the increasing difficulty of satisfying these demands from natural sources has led to a renewed interest in the

TABLE 2 Common Frog (*Rana temporaria*) and Common Toad (*Bufo bufo*): A Summary of the Reasons for Decreases in Status Given by Correspondents to the Breeding Sites Survey and Schools Survey[a,b]

	Breeding Sites Survey Replies Relate to Individual Breeding Sites		Schools Survey Replies Relate to Sites or Groups of Sites	
Reasons for Decrease	Frog	Toad	Frog	Toad
Habitat destroyed				
For development of: housing	8	4	84	25
industry	1		3	1
roads	1		7	3
playing fields	2	2	1	
Farm ponds filled in (sometimes in schools survey because of urban development)	6	2	19	4
Garden ponds filled in	2			
Sites or groups of sites drained, piped, filled in, or no longer available for unspecified reasons	7	4	37	9
TOTAL	27(31)[c]	12(37)	151(57)	42(63)
Habitat modified (dried out, silted up, overgrown, etc.)				
By natural means (including unsuitable because of adverse weather)	11	5	21	2
Because of human activity (mainly cleaning out ponds and ditches)	14	6	17	3
TOTAL	25(28)	11(34)	38(14)	5(7)

Habitat affected by pollution			
Sewage (including farm sewage)	1		6	1
Fertilizer	2		1	
Pesticide	3	1	12	2
Oil or petrol	1		14	3
Rubbish	5	3	7	3
Detergent	1		1	
Unspecified pollution	4		9	2
TOTAL	17(19)	5(16)	50(19)	11(17)
Increased mortality				
Of adults and tadpoles by animal predators	4	3	8	2
Of adults directly by humans	2			
Of adults by cars	2	1	9	5
TOTAL	8(9)	4(13)	17(7)	7(10)
Collection				
Of spawn and tadpoles by children	8		7	2
Of adults for biological experiments	3		1	
TOTAL	11(13)	0(0)	8(3)	2(3)
GRAND TOTAL	88	32	264	67

[a] Some correspondents gave more than one reason.
[b] Adopted from Cooke (1972) and printed with permission from the author and the Zoological Society of London.
[c] Figures in parentheses are the percentages of the total number of reasons.

TABLE 3 Common Frog (*Rana temporaria*) and Common Toad (*Bufo bufo*): A Summary of the Reasons for Increases in Status Given by Correspondents to the Breeding Sites Survey and Schools Survey[a,b]

Reasons for Increase	Breeding Sites Survey Replies Relate to Individual Breeding Sites		Schools Survey Replies Relate to Sites or Groups of Sites	
	Frog	Toad	Frog	Toad
Suitable habitat created				
Garden pond	17	4	14	
Others	8	5	6	1
Colonization of new site(s) after being forced out of former site(s)	3		7	1
Introduced	5	2	3	
Extending range	3		5	1
Recovery after bad winter of 1962–1963			2	1
Less predation	1		1	1
TOTAL	37	11	38	5

[a] Some correspondents gave more than one reason.
[b] Adopted from Cooke (1972); reprinted with permission from the author and the Zoological Society of London.

artificial culture of amphibians. Serious efforts are currently being made toward this end. However, costs remain high and the quantity of available cultured animals remains inadequate to meet the demand. As the cost of collecting the natural supply increases, artificial culture will certainly become economically feasible, or even necessary, to maintain the supply of healthy amphibians for at least scientific users.

3. Recommendations

We recommend that

a. A survey of the natural amphibian resources, parallel to that conducted in England (Cooke, 1972), be conducted on the North American continent to allow adequate evaluation of the supply base.

b. Increasing efforts be made to develop the culture of amphibians, both for educational and research purposes.

c. Improvements in the care and management of amphibians be instituted as early as possible to minimize the losses that now occur between the supply and the ultimate use of amphibians and to maximize the quality of research animals.

C. USERS OF AMPHIBIANS

Amphibian users are best described by the type of agency that supports their work (Table 4), the location of the research (Table 5), and the topics under investigation (Table 6). These tables were compiled from Science Information Exchange Notices of Research Projects, active as of January 1969, and were originally published in Nace (1970) where they appear with more detail and discussion.

Federal, state, and foundation agencies funded 277 projects relevant to this document (Table 4). In addition, much research on amphibians is conducted on low-budget projects. Although there is no accurate method of estimating the numbers of such projects, reviews of the index journals, including *Dissertation Abstracts*, would reveal these numbers. At the University of Michigan and at the time in question, however, for each listed project about 15 investigators were conducting research involving the use of amphibians on unlisted projects. This 1:15 ratio of listed to unlisted projects may be conservative when it is recalled that many investigators at smaller colleges not funded by nationwide granting agencies use amphibians. Thus 3,000–4,000 amphibian projects may be active, a figure, when divided into the ILAR survey on amphibian use, representing 500–600 amphibians used (received) per project. This figure seems high for

Table 4 Support of Research Using Amphibians as of January 1969[a]

	Anuran			Urodele			
Agency	Medical	Basic to Medicine	Nonmedical	Medical	Basic to Medicine	Nonmedical	Total[c] Projects
Federal	102	76	15	28	35	13	237
Percent of projects (237)	*43*	*32*	*6*	*12*	*15*	*5*	*86[d]*
State	2	–	3	–	–	4	6
Foundations	26	4	0	3	0	1	34
Percent of projects (34)	*76*	*12*	*0*	*9*	*0*	*3*	*12[d]*
TOTAL	130	80	18	31	35	18	277
Percent of projects (277)	*47*	*29*	*6*	*11*	*13*	*6*	
OBJECT TOTAL	228			84			
Percent of projects (277)	*82*			*30*			

[a] Adopted from Nace (1970).
[b] No projects were listed under more than one character.
[c] 35 projects used both anurans and urodeles.
[d] Percent of 277.

TABLE 5 Location of Research Using Amphibians as of January 1969[a]

	Academic											Nonacademic			
	Medical			Nonmedical											
Agency	Clinical	Basic	School Public Health	Vet. Med.	Zool. and Psychol.	Chemistry	Res. Center	Agric. Exp. Stn.	Museum	Dep. Un-specified		Intramur. or Institute	Hospital	Museum	Total
Federal	21	65	1	1	88	4	3	–	2	2		36	10	4	237
State	–	–	–	–	–	–	–	5	–	–		–	–	–	5
Foundations	5	18	–	–	9	–	–	–	–	–		3	–	–	35
TOTAL	26	83	1	1	97	4	3	5	2	2		39	10	4	277
Percent of projects (277)	9	30	0	0	35	1	1	2	1	1		14	4	1	
		39%						41%					19%		

[a] Adopted from Nace (1970).

TABLE 6 Topics of Research Using Amphibians as of January 1969[a]

Character	Research Topics											Total[b]
	Neuro-biology	Ion Trans. and Regul.	Pharm.-Physiol.	Cancer	Endocri-nology	Immunol. and Trans.	Develop-ment	Virology	Biochem.-Cell Phys.	Parasi-tology	Genetics	Amphib. Main.	Other	
Medical	43	40	34	15	10	10	8	4	3	3	1	–	2	173
Basic to medicine	23	5	15	–	15	4	54	1	29	4	9	2	–	160
TOTAL	66	45	49	15	25	14	62	5	32	7	10	2	2	333

[a]Adopted from Nace (1970).
[b]Totals not equal to Tables 1 and 2 because each project in "Group Applications" is recorded separately and because projects using both anurans and urodeles are recorded twice.

many projects, but may be a reasonable average when known projects using 10 times that number are considered. Also, pituitaries from 3 to 10 animals are used for the induced ovulation of a single female frog, and a conservative estimate suggests no more than 50 percent of the amphibians received for research purposes survive the period between receipt and use.

D. LABORATORY-DEFINED AMPHIBIANS

Defined laboratory amphibians (Gay, 1971; Committee on Animal Nutrition, 1972), though essential for high-quality investigations, remain largely a hope for the future. To date, the best definition (nomenclature described in Chapter III) has been attained with the axolotl (Boterenbrood, 1966). Some colonies of laboratory amphibians—such as *Xenopus* in the laboratory of Fischberg (Blackler and Fischberg, 1968) and *Pleurodeles* in the laboratories of Gallien (Guillet *et al.*, 1971) and Beetschen (1971)—do exist and meet the criteria defined for laboratory-reared and laboratory-bred animals; however, these have not attained the status of defined laboratory lines. For *Rana pipiens* efforts are now being made to establish laboratory lines (Nace *et al.*, 1965, 1966; Nace, 1968, 1970; Nace and Richards, 1969, 1972a,b,c), but animals in these lines are not yet available in significant numbers.

The importance of nomenclature cannot be overstressed during this period when defined strains of amphibians are still under initial development. Any nomenclature that is adopted should include a designation of the laboratory where the animals were developed and a key to the criteria satisfied. The geographic designation should be as precise as possible. It should specify a geographic region in sufficient detail to suggest the parent population.

It is hoped that those strains currently under development will soon become sufficiently well defined to justify publication of a standard nomenclature and growth tables as exemplified by those listings that have appeared for mice (Staats, 1972) and other laboratory animals (Poiley, 1972). Since the laboratory use of amphibians will continue to depend on animals from wild populations, terminology to designate the history of such animals must be introduced. These terms (defined in Chapter III, Section B) include wild, wild-caught nonconditioned, wild-caught conditioned, laboratory-reared, and laboratory-bred. Types of laboratory-reared and laboratory-bred populations and lines are defined in Chapter III, Section C.

II Classification and Description of Amphibians Commonly Used for Laboratory Research

A. PHYLOGENY AND CLASSIFICATION

Members of the class Amphibia are usually small animals characterized by smooth, moist, glandular skins without external scales. Fertilization is external or internal. Eggs are laid in water or moist surroundings and have some yolk and several gelatinous envelopes. Most amphibians have an aquatic larva with metamorphosis to an adult form. Respiration takes place through simple lungs, gills, skin, and buccopharyngeal membranes. The skull is flattened, with fewer bones than fishes, and has two occipital condyles. The heart is three-chambered, and 10 pairs of cranial nerves are present. Amphibians are ectothermic.

Although the class Amphibia contains approximately 3,500 species, relatively few are commonly used in laboratory studies. Only those species in which the literature indicates wide utilization in laboratory research in the United States are included below. Additional information on classification, distribution, identification, and life history may be found in *Catalogue of American Amphibians and Reptiles* (Society for the Study of Amphibians and Reptiles, 1971 *et seq.*) or in Bishop (1947), Wright and Wright (1949), Conant (1958), Savage (1961), Stebbins (1966), Goin and Goin (1971), Blair (1972), and Porter (1972). Anatomy is described in Ecker and Wiedersheim (1896).

The earliest known fossil amphibian (*Icthyostega*) was found in upper-Devonian freshwater deposits in Greenland and was probably descended from crossopterygian fishes. There are three living orders:

1. Order Apoda   Slender vermiform burrowers with compact skull, no limbs or girdles and with greatly reduced eyes that are buried in the skin. Some species are viviparous. Fertilization is internal, and the male cloaca

is modified to form a protrusible copulatory organ. Primitive genera have minute dermal scales imbedded in the skin. All living species belong to a single family, Caecilidae. Caecilians are primarily tropical forest dwellers and are not widely used in laboratory research.

2. Order Caudata   Modern salamanders or urodeles usually have slim bodies and tails, four limbs (except in Sirenidae, which has forelimbs only and is sometimes placed in a separate order, Trachystomata), and a reduction in skull bones.

3. Order Anura   Frogs and toads (salientians) are characterized by well-developed hind limbs and pelvic girdle adapted for jumping, fused head and trunk, webbed toes, 10 vertebrae and ribs reduced or absent. The term "toad" is ambiguous; it is often used for the more terrestrial anurans, but is properly applied only to the family Bufonidae.

The identification of amphibian larvae is often difficult. Orton (1952) has provided a key to the genera of tadpoles in the United States and Canada, but few developmental stage keys have been published to encompass development through metamorphosis. Gosner (1960) has proposed a simplified table for staging anurans and reference is made below to developmental tables for those few species for which standardization tables are available. Juvenile amphibians also may be difficult to identify, although reference to identification manuals, such as those listed above, may be useful.

B. DESCRIPTION OF SPECIES COMMONLY USED IN LABORATORY RESEARCH

1. Urodeles

a. *Ambystoma* ($2N = 28$; ZW ♀ - ZZ ♂, but two triploid species are known)

Species of this genus—the best known being the axolotl forms of the closely related species *Ambystoma mexicanum* and *A. tigrinum* (Smith, 1969) (Figure 1)*—are widely distributed in North America. Some of the species show a proclivity toward neoteny: The larvae grow to large size (0.18-0.30 m) (7-12 in.), retain their gills, and remain aquatic without developing all of the adult characteristics (e.g., the axolotl forms noted above). With proper alterations of the environment, most of these neotenic forms

---

*The committee gratefully acknowledges the photographic assistance of Gordon A. F. Dunn, Biomedical Graphic Communications, University of Minnesota, in making available Figures 1-12.

FIGURE 1 The tiger salamander *Ambystoma tigrinum* in the usual adult form; they also exist in a neotenic form in the western United States.

can be induced to metamorphose. Adding iodine to the water induces a faster metamorphosis since it is an essential component of thyroxine. When iodine is not present in the water, insufficient thyroxine is produced and metamorphosis is retarded.

The well-known axolotl form of *A. mexicanum* is found in Mexican mountain lakes and accounts for approximately 30 percent of urodeles used in laboratory research (Nace, 1970). *A. tigrinum* is found in the western United States, particularly in lakes in Colorado, Montana, New Mexico, Texas, and Wyoming. The taxonomic history of the Mexican axolotl has been confusing. Many generic names (including *Axolot, Axolotus, Philhydrus, Siredon,* and *Sirenodon*) have been applied to this form. These names, however, are invalid as is the emendation *Amblystoma* put forward by Agassiz in 1846. This last generic name was rejected by the International Commission of Zoological Nomenclature (Smith, 1969). Although some investigators have continued to use the incorrect "*Amblystoma*," the official name of the genus is *Ambystoma*.

The adult tiger salamander *A. tigrinum* has an olive-yellow, yellow, or black ventral surface and a dorsal ground color of brown to black with olive or brownish yellow irregular light spots that extend well down the sides. In some subspecies the dorsal markings may form light vertical bars, but the pattern is highly variable. The spotted salamander *A. maculatum* has round light yellow or orange spots on a dorsal black ground color and a slate-blue ventral surface; this species occurs from Nova Scotia and central Ontario south to Georgia and eastern Texas. Marble salamanders, *A. opacum,* are characterized by silvery irregular crossbands, gray in females and white in males, on a black background; the belly is black. This species occurs from New England to northern Florida and west to eastern Texas. Males of this genus can be recognized during the breeding season by the protuberant vent. The embryonic development of urodeles is usually staged according to Harrison's table for *A. maculatum* (see Rugh, 1965).

b. *Notophthalmus viridescens* ($2N = 22$; ZW ♀ - ZZ ♂)

The red-spotted newt (Figure 2) has a complex life history. The aquatic larvae transform into a bright to dull orange-red terrestrial eft stage and later transform into the olive-green aquatic adult. The term "newt" is generally applied to those salamanders that have a terrestrial stage ("eft") preceding the aquatic and mature adult form. In some areas, however, the eft stage may be omitted. The adult has a yellow venter with small black spots. All stages are characterized by red dorsal spots variable in number and position. Males are longer and have larger and longer vents, which may swell during the breeding season; the vent of females is small and conelike. In the past this species has been placed in the genera *Triturus* and *Diemictylus,* and reference in the literature may be found under all three names. This newt ranges from the maritime provinces of Canada to the Great Lakes southward to Florida and eastern Texas. Other species of the family Salamandridae sometimes used in laboratory studies include members of the genera *Triturus* (newts) from Europe and Asia, *Taricha* (newts) from the western United States, and *Salamandra* from Europe.

FIGURE 2 Adult red-spotted newt *Notophthalmus viridescens.*

c. *Necturus maculosus* (2N = 24; ZW ♀ - ZZ ♂)

The mudpuppy (Figure 3), like other members of the family Proteidae, is a neotenic aquatic form with gills and two pairs of gill openings. Adults range in size from 0.20 to 0.43 m (8-17 in.) and have a gray or rust-brown dorsum with indistinct scattered blue-black spots and a gray or dark belly with darker spots. Females have a simple vent; the vent of males has a wrinkled margin, crossed behind by a crescentic groove and containing two nipple-like papillae. Mudpuppies are found from western New England to Manitoba southward to northern Texas, northern Louisiana, and Tennessee.

2. Anurans

a. *Xenopus laevis* (2N = 36; ZW ♀ - ZZ ♂)

The African clawed frog, or platanna (Figure 4), like other members of the family Pipidae, is stout-bodied and big-footed. Although highly aquatic, it is found in areas where ponds may dry out during long periods of low rainfall, and under such conditions may migrate overland to other bodies of water or may aestivate in mud burrows until rains return. The common name derives from the presence of small black curved claws on the inner three toes of the hind feet. The platanna is widely distributed in Africa, south of the Sahara Desert. A significant population derived from escaped animals now exists in southern California. Females have much larger ventral flaps than those of males and are also larger in body

FIGURE 3 The mudpuppy *Necturus maculosus*—a neotenic aquatic salamander.

FIGURE 4 The African clawed frog *Xenopus laevis*.

size. Males have a black spinulose nuptial surface on the inner arms and develop fingers during the breeding season. Developmental stages of the clawed frog were described by Nieuwkoop and Faber (1956).

b. *Rana catesbeiana* ($2N = 26$; XX ♀ - XY ♂)

This species, the only anuran properly referred to by the common name "bullfrog" (Figure 5), is a member of the family Ranidae (true frogs). This family contains approximately 250 species, representatives of which are found on all continents. The bullfrog is the largest native anuran in North America, with a maximum snout-vent length of 0.20 m (8 in.). It is green above and sometimes has gray or brown irregular markings. The venter is white and is often mottled with gray. There are no dorsolateral folds on the trunk. Mature males have tympanic membranes about twice the diameter of the eye; in mature females these are approximately the same size as the eye. Males have heavy, darkly pigmented thumb pads in contrast to the more delicate, streamlined thumb of the female. This species is aquatic and is found from Nova Scotia west to Wisconsin and Nebraska southward through northern

FIGURE 5 The bullfrog *Rana catesbeiana*—the largest anuran native to North America.

Florida and Texas. It has been widely introduced elsewhere, particularly on the West Coast and in Mexico, Cuba, Japan, and Formosa.

c. *Rana grylio* ($2N = 26$; XX ♀ - XY ♂)

The pig frog (Figure 6) closely resembles the bullfrog and is sometimes supplied as "bullfrogs" by commercial dealers. This species can be distinguished from *R. catesbeiana* by the full webbing to the tips of the toes and by the fourth toe that protrudes only a little beyond the adjacent toes; in the bullfrog, the fourth toe protrudes well beyond the other toes, and the toes are less fully webbed. Sexes may be distinguished as in *R. catesbeiana*. *R. grylio* is an aquatic frog that ranges in the Coastal Plain from southern South Carolina to southeastern Texas and throughout Florida.

d. *Rana clamitans* ($2N = 26$; XX ♀ - XY ♂)

The bronze frog (Figure 7) is highly variable in coloration and may have a green to plain brown or bronze back with darker dorsal spots or markings

FIGURE 6　The pig frog *Rana grylio*, sometimes supplied as the "bullfrog" by commercial dealers, closely resembles *R. catesbeiana*.

FIGURE 7　The bronze frog *Rana clamitans*.

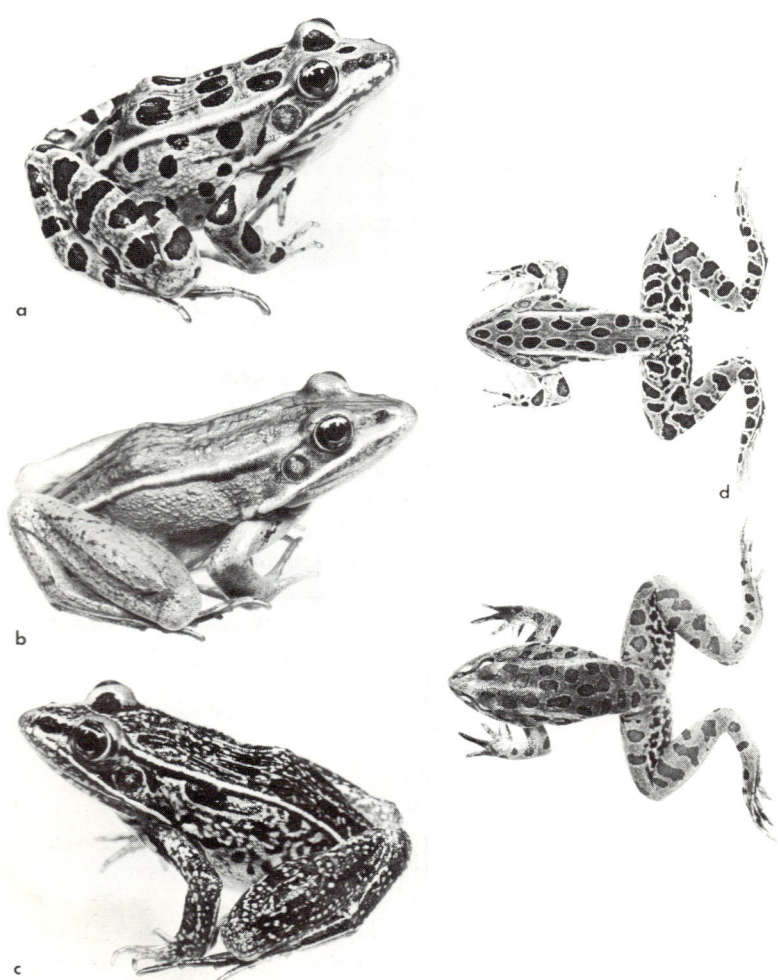

FIGURE 8 Leopard frogs, *Rana pipiens*. (a) typical member obtained from northern United States; (b) Burnsi mutant of the leopard frog, obtained from Minnesota; (c) Kandiyohi variant of the leopard frog, obtained from Minnesota; (d) *top:* a "northern frog"; *bottom:* the more pallid "Mexican frog." Note that the spots between the dorsolateral ridges of the northern frog tend to be circular in contrast to the transverse posterior spots in the corresponding area of the Mexican frog.

and a white venter that often has vermiform markings or mottling. The upper lips are sometimes green, and males often have a yellowish throat. The prominent dorsolateral ridges end on the body. Sexes may be distinguished as in *R. catesbeiana*. This species usually occurs in shallow water along streams, lakes, and springs and is abundant throughout much of its range in southeastern Canada and the eastern States.

e. *Rana pipiens* ($2N = 26$; XX ♀ - XY ♂)

The leopard frog (Figure 8) is the most widely used anuran in laboratory research (Nace, 1970). This is a highly variable species, perhaps a species complex (Littlejohn and Oldham, 1968; Pace, 1972; Brown, 1973; Platz and Platz, 1973), with wide differences in body size, coloration, and pattern. Local populations tend to differ, even between adjacent ponds. Typically, this species is brown or green with irregularly spaced dark spots on the dorsum between conspicuous dorsolateral ridges that extend to the groin. The dark spots are usually rounded and have light borders, but adjacent spots may merge. Southwestern forms tend to be light brown in color. Ordinarily, northern males have dark, thickened thumb pads, while southern males are so characterized only during the breeding season. The leopard frog is semiterrestrial and is found in shallow water habitats throughout its range; where protective cover occurs, it will often wander well away from water. The leopard frog occurs from southern Labrador to southern MacKenzie and eastern British Columbia southward throughout the United States and into Central America.

Several forms are recognized by animal dealers and laboratory investigators:

1. The "northern frogs" occur in the northern part of the range and generally fit the description often given for the subspecies *R. p. pipiens*. Adults range from 65 to 105 mm (2.5-4 in.) in snout-vent length.

2. The "southern frogs" differ from the northern leopard frog by having a light spot in the center of the tympanum, a longer pointed head, and only a few dark spots on the sides. This form occurs from southern New Jersey to the Florida Keys westward to eastern Texas and northward in the Mississippi Valley to Missouri, Illinois, and Indiana. This form has been recognized as *R. p. sphenocephala*.

3. The Rio Grande leopard frog, *R. p. berlandieri*, which occurs from southern Nebraska into Central America, is characterized by a more pallid coloration, more warty texture, more fully webbed toes, and a tendency for the posterial dorsal spots to become transverse. It differs from other forms in its adaptation to more arid environments.

4. "Mexican frogs" supplied by dealers, although often included in the above subspecies, may constitute a separate form. Such frogs have a much larger body size; have an average snout-vent length of 105 mm (4 in.), the maximum being about 130 mm (5 in.); are dark brown; and have unusually heavy hind limbs (Bagnara and Stackhouse, 1973).

5. "Burnsi" mutants occur in Minnesota and adjacent states and have no, or very few, dorsal spots.

6. "Kandiyohi" mutants appear to be mottled, because, in addition to the regular spotting pattern of this species, the ground color between the spots is invaded by dark pigment (McKinnell and Dapkus, 1973).

Both Burnsi and Kandiyohi have been incorrectly described as subspecies; instead they are single locus dominant mutants found in populations of the northern leopard frog. Melanistic forms, albinos and blues have also been reported from several locations throughout the range.

Until a few years ago no attempt was made by amphibian supply companies to distinguish Rio Grande frogs (*R. pipiens berlandieri*) from other leopard frogs, and they were supplied as "on available basis" without any great attention being paid to their distinction. Size range, pattern, and coloration, as described above, can serve as rough indications of geographic origin, but users should determine the collection point from the supplier.

Developmental stages for all species have been described by Shumway (1940) for the embryonic and early larval period and Taylor and Kollros (1946) for the larval and metamorphic period.

Life cycles of leopard frogs vary greatly between northern and southern portions of their ranges. The northern animals, which occur north of the line separating ice-free from ice-covered ponds or quiet streams in winter, lay their eggs in March or April. The progeny metamorphose 3 months later and hibernate through the winter with the adults on the bottom in water under the ice cover. Commercial collectors report movement of the frogs toward areas of water flow, presumably to regions of higher oxygen tensions. However, long periods are spent without much movement, frequently in large congregations. When water temperatures decrease from 4 to 2 °C (39 to 36 °F) in December, a shift occurs from lipid and protein utilization to carbohydrate utilization; a reverse change takes place in March when temperatures increase to 4 °C (39 °F). Under some circumstances northern frogs may hibernate below the frost line in the mud of swamps or in soft soil, but the usual behavior pattern is to migrate from swamps and fields to lakes and quiet streams, a migration that is reversed in the spring. In place of hibernation, Mexican frogs are reported to aestivate during the hot months and to breed in January or February. Frogs from intermediate portions of the range may hibernate in swamps or soft ground or be active throughout the year, depending on the ambient temperature. Accordingly, frogs for reproductive studies must be handled quite differently depending on their origin.

f. *Rana palustris* ($2N = 26$; XX ♀ - XY ♂)

The pickerel frog (Figure 9) has an overall appearance similar to some leopard frogs, but may be distinguished by the square dorsal spots arranged in two parallel rows down the back and by the bright yellow or orange color on the concealed surface of the hind legs. The pickerel frog is found in a wide variety of habitats near cool water, but like the leopard frog has a tendency to wander. This species ranges from the Canadian maritime provinces to northern South Carolina westward to southeastern Texas and Wisconsin.

FIGURE 9 Two rows of square-like spots on the dorsum and yellow limb undersurfaces distinguish this pickerel frog, *Rana palustris*, from leopard frogs similar in appearance.

g. *Rana sylvatica* (2N = 26; XX ♀ - XY ♂)

The highly secretive wood frog (Figure 10) is distinguished by the dark patches that form a mask extending from eye to ear. The general coloration varies from pink to brown or almost black. Dorsolateral folds extend along the sides to the groin. A denizen of moist woodlands, this frog may wander for considerable distances away from ponds and spends little time in the water except to breed. Wood frogs occur from the tundra of Labrador and Alaska south to the southern Appalachians with isolated populations in Kansas, Colorado, Wyoming, Idaho, and the Ozarks. They hibernate burrowed in soil under the forest litter and are usually not encountered in numbers, except during mating periods in early spring. Pollister and Moore (1937) described stages of development from fertilization to opercular development.

h. *Bufo marinus* (2N = 22; XX ♀ - XY ♂)

The giant toad (Figure 11) may exceed 0.23 m (9 in.) in snout-vent length, although apparently not in North American populations, and is characterized by deeply pitted parotid glands that extend ventrally far down the body. This very large toad just enters the United States in southern Texas,

FIGURE 10 The wood frog *Rana sylvatica* is recognized by the dark mask that extends posterior from its eyes to its ears.

FIGURE 11   The giant toad *Bufo marinus* is found in southern Texas and southern Florida.

but has been introduced into southern Florida where it is becoming fairly common. The secretions of the skin glands are toxic to many animals. Males have excrescences on the top of the first two fingers and on the inner side of the third finger and the inner palmar tubercle; these nuptial pads are particularly evident during breeding periods.

Due to introductions by man, the giant toad is now almost pantropical. The natural home of this species was probably in the tropical and subtropical New World.

Other species of *Bufo*, particularly the European *B. bufo* and the North American *B. americanus* and *B. terrestris*, have been used.

i. *Bombina orientalis* ($2N = 24$; XX ♀ - XY ♂)

The asiatic fire-bellied toad (Figure 12), a member of the family Discoglossidae, is known as *suzu kaeru* or "tinkle-bell frog" in Japanese and is a counterpart of the European fire-bellied toad, *B. bombina*. The asiatic species is found along streams in mountainous areas in Manchuria and Korea. In nature this species has a highly variable color pattern ranging from a dull solid brown to bright green dorsum or various patterning of these colors, often overlaid with black markings. The ventral surface is generally red with a variable number of irregular black markings. In juveniles, the underside is yellow and not marked with black. In laboratory-

FIGURE 12 The asiatic fire-bellied toad *Bombina orientalis* is found along mountain streams in Manchuria and Korea.

raised adults, yellow is substituted for the red, which probably reflects nutritional factors. Mature females have enlarged abdomens and rounder snouts, and the dorsal "warts" are lower in profile than in the males. Mature males have black pin-point spots (0.1–0.2 mm) (0.0039–0.0078 in.) on the apex of the rather pointed dorsal "warts," thickened forelimbs, and a black thumb pad. This species breeds in May in mountain springs or paddy fields. It hibernates in burrows or deep within talus slopes.

III Definition and Description of Experimental Amphibians

A. INTRODUCTION

Standardization of amphibians for experimental use demands that a classification be established to permit investigators to select animals most appropriate to their needs, communicate effectively with suppliers, and accurately report their data. The following standard categories are recommended:

1. Wild
2. Wild Caught
    a. Wild-Caught Nonconditioned
        (1) Wild-caught nonconditioned nontreated
        (2) Wild-caught nonconditioned treated
        (3) Wild-caught nonconditioned miscellaneous
    b. Wild-Caught Conditioned
        (1) Wild-caught conditioned larvae
        (2) Wild-caught conditioned juveniles or adults
        (3) Wild-caught conditioned miscellaneous
3. Laboratory Reared
    a. Laboratory-reared standard
    b. Laboratory-reared miscellaneous
4. Laboratory Bred
    a. Laboratory-bred standard
    b. Laboratory-bred miscellaneous

Among laboratory-reared and laboratory-bred animals the following types of populations and lines may be designated:

(1) Random mating lines
(2) Heterozygous isogenic clones
(3) Heterozygous marked lines
(4) Mutant lines
(5) Inbred lines
(6) Gynogenetic diploid lines
(7) Homozygous lines
(8) Haploid animals
(9) Polyploid animals

B. DEFINITION

1. Wild

These are pre- and postmetamorphic amphibians in nature *on which experiments are conducted in nature,* e.g., experiments on such questions as migration, population characteristics, and physiological ecology. The investigator should record the location, time, temperature, and other relevant observations concerning the collection.

2. Wild Caught

See Chapter V, Section C.2 for a special category of this classification.

a. Nonconditioned

(1) *Nontreated*

(a) General Description **Wild-caught nonconditioned nontreated** pre- or postmetamorphic amphibians refers to those collected in nature and shipped to the user with no handling or treatment other than that involved in catching, shipping to distribution points, holding between capture and sale and sorting, etc. (Gibbs *et al.*, 1971); these animals receive no disease treatment or maintenance under regularly standardized procedures. The characteristics of these animals will vary with the region from which they were captured and with the season of the year. These animals may have been maintained under a variety of environmental conditions; are usually provided no food; and, although normally shipped to buyers within a week, are often held in bulk pens for much longer periods. It is mandatory that dealers who wish recognition for meeting "standards" with respect to wild-caught amphibians specify the following:

1. species to the lowest recognized taxonomic level,
2. geographic origin of the parent known to the closest possible geographic unit,

3. date and method (see Chapter III, Section C) of reproduction known and the date of metamorphosis recorded to within 1 month,
4. method and period of holding, and
5. the environmental conditions to which the animals were exposed from fertilization to shipment to the user and, if possible, the environmental conditions during shipment.

A subdivision of this classification is "northern frogs" or *R. pipiens* capable of being ovulated during the winter months. Generally, they are collected north of the line separating ice-free from ice-covered ponds and streams. They may also occur at altitudes above the ice line. In contrast, "southern frogs" are *R. pipiens* that cannot be ovulated in this season and are collected south of the ice line. Dealers must be most cautious concerning admixture of these animals; their reproductive cycle is not the only physiological difference between them.

Shipments of amphibians may contain variants. Thus, while northern collections of *R. pipiens,* with the exception of areas in Minnesota and part of Wisconsin, are reasonably uniform, some collections of *R. p. pipiens* may contain the Kandiyohi or Burnsi mutants. However, neither of these mutants constitutes more than 5 percent of the population, even in areas where they are most abundant.

Other variants may also occur, for example, admixture of frogs belonging to different segments of the *R. pipiens* complex (Brown, 1973), or from populations with lower or higher incidence of the Lucké renal adenocarcinoma. Although *R. catesbeiana, R. clamitans, R. pipiens, R. palustris,* and *R. sylvatica* occur in the same areas where northern collections are obtained, dealers seldom include these species in shipments of *R. pipiens.* Caution should be exercised, however, since rapid sorting of animals occasionally results in admixtures of *R. catesbeiana* and *R. clamitans* and of *R. pipiens, R. palustris,* and juvenile *R. clamitans.*

Southern collections include frogs from the southern states of the United States and from Mexico. Such collections may contain mixtures of *R. p. pipiens* and of *R. p. sphenocephala* if made in the central states and of *R. p. berlandieri* if from the southern states through Mexico and Central America. The systematics of these species, however, has not yet been fully resolved.

Among the other species, the nature of variations within collections has not been well defined. Physiological variants must be expected where the species range extends longitudinally. Among other physiological variants sexually "differentiated" and "undifferentiated" populations of *R. catesbeiana* have been identified (Witschi, 1930), but the geographic coordinates of these populations have not been defined, which exemplifies the need for specification of geographic origin.

These **wild-caught nonconditioned nontreated** animals must fulfill other criteria before they can be placed under a higher classification.

(b) Embryos or Larvae   Embryos or larvae are, of course, premetamorphic stages. There is no opportunity to treat or condition wild-caught embryos, commonly known as egg clutches, or young larvae prior to shipment. Thus, they are classified under this category. Larvae held for longer periods may qualify for classification under another category.

(2) *Treated*   **Wild-caught nonconditioned treated** are animals that meet all of the criteria for the wild-caught nonconditioned nontreated classification and usually are provided some food prior to use or shipment to a buyer. A unique character of this classification is that attempts have been made to treat the animals for disease or parasites, although it must be recognized that standard treatments remain undefined. It is important that the investigator know the treatments to which the animals have been exposed. Normally, these animals are shipped within a few days after being received by the supplier but may, in fact, be held for much longer periods. Dealers, in addition to meeting the *five* specifications listed above, must also specify

6. treatments to which the animals have been exposed.

(3) *Miscellaneous*   This category includes treated or nontreated embryos, larvae or adults for which two or more of the specifications listed under nontreated are not fulfilled.

b. Conditioned

Wild-caught conditioned amphibians not only fulfill the *six* criteria established for nonconditioned animals [see Section III.B.2.a(1),(2)] but also meet the specification that records are available, indicating:

7. their treatment, if any, and length of exposure to each laboratory environment,
8. pathological and general physiological state (e.g., whether or not in hibernation, known diseases, feeding behavior, activity), and
9. approximate age (if known).

(1) *Larvae*   **Wild-caught conditioned larvae** are wild-caught amphibian larvae or larvae from eggs collected in the wild. They should be

maintained under standard laboratory conditions for a period sufficient to demonstrate that their mortality rate is not appreciably greater than is normal for laboratory-reared representatives of the species or mutant in question. If the larvae are maintained through metamorphosis, they should fulfill the *nine* requirements for conditioned juveniles and adults

(2) *Juveniles or Adults* **Wild-caught conditioned juveniles or adults** are wild-caught amphibians maintained under standard laboratory conditions for a sufficient time to demonstrate for the species in question the absence of symptoms of disease or physiological disorders. For example, *R. catesbeiana* adults should be held for a minimum of 6 weeks, during the last 2 weeks no symptoms of disease or physiological disorder should be in evidence.

Newly metamorphosed *R. catesbeiana* should be maintained under standard conditions for a minimum of 10 weeks. This assures that they have survived the 2-month period of high immediate postmetamorphic mortality when disease symptoms are not easily recognized and death is too rapid for medication to be administered.

(3) *Miscellaneous* Wild-caught amphibians maintained under the conditions needed to assure freedom from symptoms of disease or physiological disorder, but lacking in the other requirements for conditioned amphibians, are classified as wild-caught conditioned miscellaneous.

3. Laboratory Reared

a. Standard

Laboratory-reared standard are amphibians reproduced in the laboratory with at least one wild-caught parent or have been fed living food items collected from nature or living food items exposed to intermediate parasite hosts [see Chapter V, Section C.2 and Chapter VI, Section B.1.b(2)]. The offspring must fulfill the *nine* criteria required of **wild-caught conditioned amphibians**. The parents may be collected at any stage of development. This classification recognizes the possibility of disease or parasite transmission from a parent or from food exposed to intermediate hosts.

b. Miscellaneous

Laboratory-reared miscellaneous amphibians meet the requirements for **laboratory-reared standard** with at least one field-collected parent but lack such pertinent information for proper classification as geographical origin of parents, laboratory conditions, disease treatment, and age (see Chapter III, Section B.2).

4. Laboratory Bred

a. Standard

Laboratory-bred standard amphibians are those produced by reproductive events that did not occur in nature and whose parents were not field collected; they are fed processed food items or living food items bred under laboratory conditions isolated from the amphibian population. Thus, **laboratory-bred** amphibians must fulfill all the criteria required of **laboratory-reared** amphibians and must be at least of the F_2 generation.

Laboratory-bred standard amphibians are free of parasites and symbionts that require intermediate hosts; they may, however, possess those parasites and symbionts that are regularly transmitted vertically and without intermediate hosts.

b. Miscellaneous

Laboratory-bred miscellaneous amphibians are those fulfilling all **laboratory-bred** requirements except for maintenance on a diet of food isolated from the natural environment. For example, in some areas, *R. catesbeiana* is maintained in the laboratory predominantly on a diet of living fish from outdoor ponds [see Chapter VI, Section B.1.b(2)]. Suppliers of **laboratory-bred** *R. catesbeiana* are reminded that this classification requires documentation of the types of food being used.

C. DESCRIPTION OF LABORATORY-REARED AND LABORATORY-BRED AMPHIBIANS

1. Types of Populations and Lines

Amphibians are unique in that no other laboratory animals are capable of reproduction by as many significantly different procedures. They can be reproduced by natural biparental mating, natural parthenogenesis, artificial insemination, various types of artificial parthenogenesis, or by nuclear transplantation. Since each procedure has significantly different genetic consequences, the method of reproduction must be specified when defining laboratory-reared or laboratory-bred amphibians (Asher, 1970, in press a,b; Nace *et al.*, 1970; Asher and Nace, 1971). The following is an outline of the various lines that may result from the application of these several methods of reproduction. It is simplified in that terms are not suggested for many of the types of genealogies that could result from combinations of the several methods of reproduction.

a. Random Mating Lines

Random mating lines refer to genealogies resulting from the bisexual mating of random animals within a population. Reproduction may be by natural mating in season, by hormonally induced mating, or by artificial insemination in or out of season. As a first approximation the progeny of such matings will be as genetically heterogeneous as the population from which the parents were chosen.

b. Heterozygous Isogenic Clones

Groups of animals produced from wild-caught animals or random mating lines by the technique of nuclear transplantation (Chapter VII, Section A.9) comprise heterozygous isogenic clones if the nuclei used in the transplantations are all from a single individual. Each member of such a clone is genetically identical to its clone mates. Because they are isogenic, each will accept grafts from the others (Volpe and McKinnell, 1966). However, since the nuclear transfer frogs are heterozygous, biparental progeny of members of a clone are as genetically heterogeneous as the progeny of matings within any set of identical siblings.

Clones may be produced by the nuclear transplantation technique from animals in any of the populations or lines described below. In such cases, clone members with the genetic properties of the parental line are produced. These may be more or less heterozygous depending on the state of the parental line.

c. Heterozygous Marked Lines

Increasing numbers of mutations in amphibians are being described. Many produce characteristics that are significant to investigators (Briggs, in press; Malacinski and Brothers, in press). Some of the phenotypes are apparent from external examination; others are biochemical or developmental mutants that can only be detected by special techniques. It is advantageous to link biochemical and developmental mutants to the externally visible mutants to allow ready laboratory manipulation. Consequently, in certain lines, pigmentation or pattern mutations are being selected without regard to the status of the remaining genome. Such lines are characterized as containing heterozygous marked animals into which less evident mutations can subsequently be introduced.

d. Mutant Lines

Mutant lines may include heterozygous marked lines or lines of animals selected for any specific mutations with the remainder of the genome specified to varying degrees.

e. Inbred Lines

Inbred lines may arise from any of the previously defined lines and are characterized by varying degrees of genetic homogeneity. They may be produced either by a sequence of selected biparental matings or by any of the various techniques of parthenogenesis.

Sequential biparental matings of amphibians yield inbred lines comparable to those common to inbreeding within other kinds of organism. Brother-sister or parent-child matings are utilized to minimize the genetic diversity among the progeny. A sufficient number of generations of such breedings can result in specified degrees of genetic homozygosity with the limitations imposed by low viability, genetic drift, and gene fixation that are well known from inbreeding within other organisms.

f. Gynogenetic Diploid Lines

Gynogenetic diploid lines are special inbred lines produced by a modification of parthenogenesis, which, as generally applied, is genetically equivalent to fertilizing the egg with its own second polar body (see Chapter VII, Sections A.5 and A.6). This technique does not immediately produce animals homozygous at all gene loci. The first generation is homozygous for those loci located close to the kinetochore, but crossing-over results in increasing heterozygosity as the gene-kinetochore distance increases. Nevertheless, three generations of gynogenetic reproduction of *R. pipiens* is genetically equivalent to 22 generations of biparental inbreeding in mice (Nace *et al.*, 1970; Asher and Nace, 1971). The limitations of low viability, genetic drift, and gene fixation apply.

g. Homozygous Lines

Homozygous lines refer to those lines in which most or all gene loci are in the homozygous state. "Standard" homozygous lines refers to the former; "absolute" homozygous lines refers to the latter.

Lines produced by three or more generations of diploid gynogenesis may be considered **standard homozygous lines** with the limitations noted above. **Absolute homozygous lines** of several types may be produced, e.g., gynogenetic, androgenetic, or nuclear transplant recipients. In gynogenetic lines, the genome is totally that of the female progenitor; in androgenetic lines, it is totally that of the male. The process of producing gynogenetic diploids is initiated but without the step that results in the retention of the second polar body. Such animals would be haploid if left untreated. At the time of first cleavage, such eggs are exposed to hydrostatic pressure of 5,000 psi ($\sim 3.5 \times 10^7$ Pa). This suppresses cytoplasmic division but allows nuclear division and the reconstitution of the diploid state from the original haploid set of maternal chromosomes.

In the case of androgenetic homozygous animals, the female pronucleus of artificially inseminated eggs is removed before union with the male pronucleus. The diploid state is reconstituted in the progeny from the haploid paternal set of chromosomes by exposure to high pressure at the time of first cleavage, as in the case of the gynogenetic animals. Animals in absolute homozygous lines produced from animals other than those in standard homozygous lines show extremely low viability. Those from standard homozygous lines show higher viability because of the selection involved in the production of the parental lines.

Homozygous diploids may be produced by transplanting haploid nuclei to enucleated eggs (see Chapter VII, Section A.9). A delay in cytokinesis occurs spontaneously in only a few eggs receiving transplanted nuclei. The delay in cytokinesis can also be produced by pressure treatment. The haploid nucleus undergoes mitosis, and the daughter nuclei fuse to produce a homozygous diploid. Normal cytokinesis follows after the delay of one cleavage interval. Nuclear transfer homozygous diploids, as absolute homozygous lines, have low viability (Subtelny, 1958).

h. Haploid Animals

Death as a result of the haploid syndrome, which usually expresses itself at early morphogenesis (tailbud stage), prevents the development of haploid lines (Porter, 1939). Haploid individuals, however, may be produced by the same technique used for the production of gynogenetic or androgenetic diploid animals, except that the step resulting in the retention of the second polar body or suppression of the first cleavage is omitted. In spite of their poor viability, such haploid animals are potentially useful as sources of haploid lines of cells for tissue culture (Freed and Mezger-Freed, 1970) or other experimental procedures.

i. Polyploid Animals

Polyploid animals and polyploid lines of almost any specified type may be produced by the combination of several techniques (Kawamura and Nishioka, 1963, 1967, 1972, 1973). Thus, normal biparental mating followed by retention of the second polar body, as in the production of gynogenetic diploids, results in triploid animals with two chromosomal sets of maternal origin (Dasgupta, 1962). The production of diploid animals either by gynogenetic techniques or by normal biparental mating followed by the suppression of the first cleavage division results in the production of tetraploids. In the second case, two sets of chromosomes are of maternal origin and two are of paternal origin. By using this technique with diploid, triploid, or tetraploid parents, many other varieties of ploidy are possible (Fankhauser, 1945). Frogs of several ploidy classes are routinely observed in nuclear transplantation experiments (McKinnell, 1964) (see Chapter VII, Section A.9).

2. Sex Determination and Its Manipulation

The sex determination of wild or wild-caught amphibians does not constitute a problem because it usually follows expectations. However, among laboratory-reared or laboratory-bred amphibians unusual sex ratios may be observed. Because sex determination may be a critical variable to the investigator, sex determination must be considered here.

Sex determination in amphibians follows either the XX-XY or ZW-ZZ form of genetic control (see Chapter II, Section B). The former is typical of Ranidae; the latter is typical of the urodeles. Note that *Xenopus* differs from the Ranidae as it is of the ZW-ZZ type (see Chapter II, Section B.2.a). Thus it would be expected that in the case of Ranidae, diploid gynogenesis, in which the genome is totally of maternal origin, should result in the production of 100 percent females. In actual practice, however, it has been found (Richards and Nace, unpublished) that males may be produced in unexpectedly high frequencies, both in certain gynogenetic reproductions (3.6 ♀ : 1 ♂ among 1,234 progeny of 106 females in 4 years) and in the biparental matings of frogs from different geographical areas (1 ♀ : 17.4 ♂ from northern females × Mexican amelanoid males). As a result, it is the practice in some laboratories to administer 50 µg/liter of β-estradiol or testosterone to larval stages to control the numbers of males and females needed for breeding stock. Thus, animals may be phenotypically female but genetically male or vice versa. Untreated biparental progeny of such sex-reversed animals develop in accordance with the sex specified by their genetic composition.

The above facts, plus the normal lability of sex determination in certain species of amphibians, should warn the user to exercise caution in interpretations of experimental results that may vary as a consequence of a disparity between the genetic and phenotypic sex of the animals in the experimental groups.

3. Species of Laboratory-Reared or Laboratory-Bred Amphibians Available in the United States

The number of species of amphibians that are available as laboratory-reared or laboratory-bred animals remains limited. Most extensively bred is the *A. mexicanum,* the axolotl. Laboratory-bred *X. laevis* are now available in a number of laboratories. *R. pipiens* and *B. orientalis* have been bred and are becoming available.

a. Anurans

(1) *R. pipiens* Random mating and heterozygous marked lines are available. Other lines are in production or may be started by the investigator using the random mating or marked lines. The heterozygous marked lines are being developed using the best known mutants of *R. pipiens* [e.g., burnsi (Moore, 1942), kandiyohi (Volpe, 1955), and amelanoid (albino) (Smith-Gill *et al.*, 1972)]. The melanoid mutant (Richards *et al.*, 1969) is still under analysis but may soon yield a marked line. A few biochemical and other mutants that require special testing are also being produced as mutant lines. Inbred lines should be available within several years. At the time of this writing, arrangements can be made with the Amphibian Facility at the University of Michigan to obtain some animals from these lines.

(2) *R. catesbeiana* "Differentiated race"-sex stable, "undifferentiated race"-reversal from female to male is common after metamorphosis (Witschi, 1930; Hsu and Liang, 1970). Laboratory-reared and laboratory-bred animals are now being produced by Dudley D. Culley (School of Forestry and Wildlife Management, Louisiana State University, Baton Rouge, Louisiana) and other suppliers are initiating production efforts (Nace *et al.*, 1971).

(3) *B. orientalis* Random mating lines of this species are now available as laboratory-bred animals from the Amphibian Facility at the University of Michigan or the Laboratory for Amphibian Biology at the University of Hiroshima.

(4) *X. laevis* Animals of this species can be obtained from a variety of sources. None routinely provide laboratory-bred animals, although these are available on occasion from the Amphibian Facility at the University of Michigan and from other private investigators in the United States. The largest colony is in the laboratory of M. Fischberg of the University of

Geneva, who maintains random mating and mutant lines of *X. laevis* and a number of other *Xenopus* species.

A. mexicanum   The Mexican axolotl is available from a number of laboratories. However, the Indiana University Axolotl Colony of R. R.

b. Urodeles

A. mexicanum   The Mexican axolotl is available from a number of laboratories. However, the Indiana University Axolotl Colony of R. R. Humphrey possesses these in largest number and has developed the largest inventory of mutants (Briggs, in press; Malacinski and Brothers, in press). Heterozygous marked lines, mutant lines, and inbred lines have been developed.

No other species of urodeles are normally available as laboratory-bred animals.

IV Sources

A. SELECTION

Guidelines for the selection of species for specific investigations are beyond the objectives of this document. However, certain general principles should be borne in mind. In order of availability or accessibility to the investigator, one may list wild-caught, wild, laboratory-reared, and laboratory-bred animals. Although the quality of the data may be in proportion to the degree of definition of the animals (see Chapter III), the current state of the art and of supply is such that the investigator should choose the type of animal that is most abundant, yet adequate for his investigation. If simply a live frog is sufficient and if parasite or genetic control is not necessary, choose wild or wild-caught animals, at least until the supply situation has improved.

Perhaps the most important criterion in a selection process is to choose a supplier whose animals are reliable. "Sources of Amphibians for Research" (Nace *et al.*, 1971) lists sources, species, and geographic origins for approximately 150 species of anurans and urodeles. Most of these species are only available from other investigators and on a seasonal or "on hand" basis. The list also includes the several large dealers that supply over 80 percent of the amphibians used in research. The investigator should learn to know the seasonal characteristics of his animal and the resources and reliability of his supplier (Emmons, 1973). He should determine whether the supplier is a collector or whether the supplier obtains his animals from a primary supplier. The adequacy of the supplier's facilities must be determined. Those who expect to use many animals should visit the facilities of the suppliers.

Endangered amphibian species, which are not available from commercial suppliers, are listed in Appendix A. Reputable users and dealers do not undertake the use or supply of endangered species.

B. DEALER CARE

An important responsibility of the dealer is to reduce the stress on the animals in his charge for the period he holds them in his establishment. This presents problems when dealing with animals throughout the year and from different geographical origins, as their physiological state is highly dependent on season and origin (see Chapters II, V, and VI). It is useful for the dealer to know the objectives of his users: For example, is the user engaged in studies involving reproduction, cellular physiology, or organ physiology? Dealers must establish those procedures in their plants that will meet the requirements for the standards of animals they choose to supply (see Chapter III). For example, hibernating northern frogs can be held in dealer's facilities if water quality (see Chapter V, Section B.2.a) and other conditions (see Chapter V, Section C.5 and Chapter VI, Section B.1) are controlled. In the past, hibernating frogs have been kept by many dealers in small ponds that freeze over in winter and are supplied by untreated water from underground springs. In general, this is a poor practice; the water is often oxygen deficient and too warm for optimal hibernation conditions.

Field studies have indicated that the hibernating frog in nature is extremely sensitive to its surroundings, adjusting repeatedly throughout the winter to changes in the environment. The two most significant environmental variables appear to be temperature range and oxygen availability, and it is not uncommon for the hibernating frog to move several hundred yards from one location in a lake or stream to another to optimize these two factors.

Interest has been shown by the research community in wild-caught conditioned animals. Commercial dealers are attempting to meet this demand for *R. catesbeiana* and in a more limited way for *R. pipiens*. The latter is limited because *R. pipiens* is more difficult to feed artificially, and the established price structure restricts the introduction of the necessary technology.

C. ORDERING, SHIPPING, AND RECEIVING

1. Ordering

The user is responsible for ordering animals in such a manner that the dealer can reasonably be expected to meet his desires.

The user must be specific in stating requirements as to size, geographic origin, nature of the research, and level of definition required (see Chapter III, Section B). Supply limitations must be recognized. For example, in dealers' pens most *R. pipiens* over 90 mm (3.5 in.) in body length are female. Therefore, orders for "105 mm (4 in.) grassfrogs (half male and half female)" are frustrating to the dealer, if not impossible, to fill.

The purchasing department should include the name of the user on the order and, if possible, reference to a previous order for similar animals. This permits the dealer to maintain uniformity in shipments.

Order amphibians by "use date," rather than "shipping date." Leave the time and method of shipment to the supplier. Though dealers limit their guarantee to live arrival, reputable dealers attempt to ship under optimal circumstances to ensure healthy arrival.

Do not order amphibians *for delivery* on Monday mornings. This generally necessitates a weekend holding period. Upon receipt of the animals, it is important that they receive prompt attention as described below.

2. Shipping

Shipping is the responsibility of the dealer who must meet his guarantee of live arrival. Any procedures that interfere with an amphibian's ability to regulate its body temperature will result in the loss of animals. Amphibians, as ectotherms, regulate their temperature by selecting an appropriate habitat or posture—processes impossible in transit. The supplier must provide the correct environment for shipping. The most difficult times for shipping ranids are periods of temperature and humidity extremes. Inadequate cooling in the summer results in what is known in the jargon of amphibian dealers as "hot frogs"—the state of being hyperactive to the point of convulsion.

Although frogs and salamanders may be shipped in boxes containing sphagnum moss or other protective material, damage caused by improper handling *en route* may occur (Gibbs *et al.,* 1971). Aquatic forms such as tadpoles, axolotls, and *Xenopus* may be shipped in thermos jugs or plastic bags in insulated containers. The numbers of animals that can be shipped per unit volume is critical and varies greatly with species and size. In the summer, the containers should include ice to minimize activity and sealed containers should also include oxygen to facilitate respiration.

A given shipping container should contain animals of only one species. In some cases sexes should also be separated, e.g., amplexing animals. Hibernating animals may be shipped submerged in cold water 2-4 °C (36-39 °F), nonhibernating animals in well-ventilated cartons. Before shipment, food should be withheld from amphibians for a sufficient time to prevent contamination of the shipping container. This is particularly true of aquatic forms such as *Xenopus.*

Most amphibian deliveries to laboratory facilities should be either by Air Parcel Post Special Delivery or Air Express.

Accompanying the shipment, suppliers should include information as follows if the amphibians are to be used for research:

- Simple directions for handling and holding the animals after their arrival with references to more extensive sources of information such as this document and
- The standard met by the animals and the information specified by the criteria for the standard (see Chapter III).

3. Receiving

The user is responsible for receiving the animals in such a manner that the specifications in his order and the efforts of the dealer are not nullified. The details of procedures to be followed upon receipt of amphibians depends on the species. However, certain general principles can be stated. Unless the user is prepared to develop suitable facilities and meet demanding personnel requirements for the long-term care of amphibians, they should not be held for more than a few days. If long-term storage or observation is necessary, prepare the required facilities prior to placing the order for animals (see Chapters V and VI). It is important to realize that seasonal variations determine the hardiness and the nature of the care required by most amphibians. Thus, northern *R. pipiens* that have spawned are not as hardy as animals collected in the fall of the year after feeding actively throughout the summer prior to hibernation.

a. Aquatic Amphibians

The aquatic animals such as axolotls, *Xenopus*, and tadpoles should be transferred immediately from the shipping containers to previously prepared containers. Last minute cleaning of spare containers often results in shoddy cleaning and inadequate rinsing to remove soap or detergent!

Do not discard the water in which the animals were received. The rapid transfer of animals to water that differs radically in quality may result in severe shock or death. As a regular practice, animals should be conditioned to the water in the laboratory by slow dilution of the transport water. Strain it through gauze and use it in the clean containers. Save half of this water in case the following steps cause the animals to show discomfort. If this water in which you will place the animals is warmer than the temperatures specified in Chapter VI for the species in question, add ice made from dechlorinated laboratory water (see Chapter V, Section B.2.a), but do not expose the animals to temperature changes greater than about 1 °C (2 °F) per hour. After the animals become quiescent and give signs of adapting to the new environment, add no more than one-fourth the volume of laboratory water. If no distress is obvious after several hours, more fresh water may be added. Repeat these steps until the animals have adjusted to your water supply. If they show distress at some step, return them to the previous water mixture in which they did not show distress and prepare new water from a different source.

Water quality is only one aspect of their new environment. In addition, container size, shape, and transparency must be considered. Individuals with obvious stress or damage should be isolated and treated as discussed in Chapter IX, Section B.2. After acclimatization, normal care procedures may be followed.

b. Terrestrial Amphibians

Because of their shape and posture, terrestrial urodeles ship quite well unless seriously overheated or dehydrated. They may be handled in much the same way as the aquatic animals, i.e., they should be moved to permanent housing through a series of gradual temperature steps.

To avoid confusion and damage to terrestrial anurans, the shipping box should be emptied into an enclosure such as a large plastic garbage can containing 0.15–0.25 m (6–10 in.) of water in order to deny the frogs a surface from which they can jump. Wash them free of packing material and separate the apparently healthy from obviously sick or damaged animals. In the winter, the wash water should be cooled to the temperature of the animals; in the summer, water at tap water temperature may be used. Do *not* wash them vigorously under a full flow of water from a tap or hose. Plastic vegetable crispers with slide-on lids are good isolation containers. The containers should be filled with water to about shoulder height for leopard frogs; toads need only a damp bottom.

From October through March and depending on the research objectives, wild-caught nonconditioned animals from north of the ice line may be returned to hibernation. This may be done by taking them from shipping temperature to hibernation temperature through a series of 9–12 h at each of several intermediate temperatures with each step being 3–5 °C (6.4–10.7 °F). Alternatively, if 20 or more gallons in the hibernaculum are used (see Chapter V, Sections C.5 and 6.a), the animals may be placed in the water at room temperature. They will acclimate as the water cools when the container is placed in the coldroom. This procedure should be followed within hours after receipt of the animals if they are to be used for their eggs (see Chapter II, Section B.2.e). *Never use* this chilling procedure with summer or Mexican frogs.

Adult *R. pipiens* to be used at room temperature, or whose eggs are not

		JAN	FEB	MAR	APR	MAY	JUN	JUL	AUG	SEP	OCT	NOV	DEC
1.	Connecticut												
	Maine												
	Massachusetts												
	New Hampshire												
	Vermont												
2.	Delaware	1			30								
	Maryland												
	New Jersey												
	New York						15				1		
	Rhode Island												
3.	Michigan					29						16	
	Minnesota				1	15							
	North Dakota				1	31							
	South Dakota					1					16		
	Wisconsin					1							
4.	Idaho												
	Montana												
	Oregon												
	Washington												
	Wyoming												
5.	Kentucky												
	North Carolina												
	Tennessee												
	Virginia												
6.	Illinois						14			1			
	Indiana					1	14						
	Ohio					2	14						
	Pennsylvania						30					2	
	West Virginia						12		1				
7.	Colorado							31			1		
	Iowa						1						1
	Kansas						30				1		
	Missouri						29						1
	Nebraska						30					2	
8.	Alabama												
	Florida												
	Georgia												
	Mississippi												
	South Carolina												
9.	Arkansas	1			15								
	Louisiana				1	31							
	New Mexico							31		1			
	Oklahoma												
	Texas												
10.	Arizona					31							1
	California					31							1
	Nevada												
	Utah												
11.	Alaska												
	Hawaii												

FIGURE 13 Amphibian closed season regulations. Hatched time spaces represent the closed seasons. Numbers in the squares refer to dates. The states are listed by tiers: northern, 1–4; central, 5–7; southern, 8–10; and other. Data on the Canadian provinces may be found in Appendix B. Data assembled with the assistance of Stafford Cox, student of environmental law, University of Michigan.

needed, may be maintained at room temperature at any season (see Chapter V, Section C.4). They must be fed, however, if they are to be kept more than a few days.

D. Legal Aspects

It is incumbent upon both dealers and users of amphibians to be aware of and abide by the laws established to regulate their use. At the federal level the Endangered Species Conservation Act provides for the conservation, protection, and propagation of any wild mammal, fish, wild bird, *amphibian*, reptile, mollusk, or crustacean threatened with extinction or likely within the future to become threatened with extinction. It further directs that, insofar as is practicable and consistent with the primary purposes of bureaus, agencies, and services, all federal departments and agencies shall exert their authorities in furtherance of this Act. The secretaries involved determine, at their discretion, the endangered species based on the best scientific and commercial data available; and after consultation with the affected states, interested persons, and organizations, the secretaries publish the endangered species list in the *Federal Register*. Upon the petition of an interested person, the secretaries conduct a review of any listed or unlisted species proposed to be removed or added to the list. The evidence required for such a review pertains to any one of the following factors: the destruction, drastic modification, or severe curtailment of habitat; overutilization for commercial, sporting, scientific, or educational purposes; the effects of disease or predation; the inadequacy of existing regulatory mechanisms; or other natural or man-made factors affecting continual existence of the species in question. With respect to amphibians, the documentation and quantification of these factors is not defined. A listing of amphibians judged to be endangered is to be found in Appendix A.

State regulations concerning amphibians are highly variable as seen in Appendix B. These regulations are summarized in Figure 13, which shows the closed seasons for "frogs" in each state. In this figure, the states are assembled by geographic regions. Figure 13 and Appendix B do not show that collection for scientific purposes is exempt or given special treatment by the law in some states. Caution should be exercised in interpreting Figure 13 as state regulations affecting frogs are currently under review in many states.

V Physical Facilities

A. RELATION TO NATURAL HABITAT

Two choices are offered when the decision is made to use a wild living animal in the laboratory and to develop it as an animal adapted to the laboratory. The first choice is to provide a laboratory habitat that is, in as many particulars as possible, a close mimic of the natural habitat. The second is to develop a laboratory habitat in which the animals live and which is at the same time compatible with the laboratory environment, minimizes labor and material costs, and may be managed on the basis of principles already familiar to animal husbandry personnel.

Many amphibian hobbyists have chosen the first alternative. With animals from a known and limited geographic territory, it is possible to construct reasonable facsimiles of a particular natural habitat in the laboratory. Usually, this is expensive, either in materials (container fabrication, water, soil, plants, inserted objects), or in labor, or both. This is difficult to do, however, when designing quarters suitable for any representatives of a species whose range places them in many quite different habitats. Also, such naturalistic quarters usually will not accommodate animals in densities that significantly exceed those found in nature.

For these reasons the second alternative was chosen, and facilities are described that are only now under critical test (Boterenbrood, 1966; Frazer, 1966). Many of the details will change as experience broadens. It is asked that experiences that may aid in satisfying the requirements of this second choice be communicated to the Institute of Laboratory Animal Resources.

The principles for the second alternative are as follows:

• No matter how varied the environments within which a given species is found, each of those environments contains common features that make

it possible for the species in question to exploit it. Only these common and essential features need or should be incorporated in the design of laboratory quarters for the species.
- Laboratory quarters and management protocols for the species in question should not require unusual expenses and should not require training that is totally different from the typical training experience of personnel in animal facilities.

B. THE AMPHIBIAN QUARTERS

1. General Description

Amphibians should *not* share quarters with mammals or birds, but may be in rooms with aquaria containing fish or other aquatic forms. The high humidity of amphibian quarters and the optimal temperatures for these ectotherms are not usually compatible with the requirements for endotherms.

The work area requirements for an amphibian unit remain the same regardless of the size of the unit. The following describes a unit suitable for the maintenance of several thousand or more animals of several species. Smaller units should contain equivalent work and animal areas even though these areas are not in separate rooms. The major difference between smaller and larger units is that smaller units may not have as many options for maintaining animals at a variety of temperatures and under various lighting regimens. Facilities for dealers will differ in size and in the proportion of areas for temporary holding as distinct from long-term animal culture.

As noted below, the amphibian unit should be provided with inflow and outflow entrances to a suite of rooms or functional areas that include:

- The animal rooms
- Breeding area
- Isolation quarters
- General laboratory area
- Examination and autopsy area
- Insectarium
- General service area
- Storage area
- Information control area
- Office area

a. Isolation Quarters

Isolation quarters for newly arrived animals are advisable. Because of the difficulties inherent in regulating temperatures for different species, specific areas within a room should be set aside for new arrivals. The new arrivals should be kept relatively isolated from other animals in the room. Since amphibians do not introduce as much contamination into the environment as do mammals and birds, their physical isolation is satisfied by placing them on the lowest cage rack level; in this position, the effluent water from their containers will flow directly into the drain.

Isolation and acclimatization for an incubation period is desirable, particularly if juveniles or adults are brought into the laboratory. Isolation not only provides disease protection to existing laboratory stocks but also gives the new arrivals a chance to adapt to their environment with a minimum of disturbance. These animals often will not feed for several days and will be easily startled by routine care activities. Minimizing activity around the isolation quarters for several days will reduce mortality.

If symptoms of disease appear, treatment and container care as discussed in Chapter IX should be followed. Diseased amphibians in isolation should not be incorporated into existing stocks until the disease has been controlled for at least 2 weeks.

b. Heating, Ventilation, and Size Specifications for Rooms

Because of the several temperature requirements for different amphibians, both during hibernation and periods of activity, the unit should be provided with animal rooms at different temperatures. This includes rooms for adult and larval amphibians and for insect culture; the size of each will depend on the size and objectives of the animal colony. To facilitate servicing, however, these should be at least "walk-in" size.

For hibernation of northern species, temperatures of 0-2 °C (32-35 °F) and 2-4 °C (35-39 °F) must be available. This requirement can be met in two ways: A sufficiently large room held at approximately 18 °C (64 °F) could contain fiberglass circulating refrigeration units to attain the temperature desired (see Section C.5). Alternatively, two hibernation rooms— one maintained at 0-2 °C (32-35 °F) and the other at 2-4 °C (35-39 °F)— could be used. The latter would have the advantage of allowing the use of hibernation containers for smaller groups of animals (see Section C.6.a). For hibernating animals from the intermediate geographic ranges, for conditioning northern animals to hibernation, and for maintaining certain of the urodeles, a room maintained at 7-12 °C (45-55 °F) is advisable.

For certain larvae and the adults of other urodeles, a room of 18-20 °C (65-68 °F) is advisable. For axolotls and for *R. pipiens* in the process of acclimatization to hibernation, a room at 20-22 °C (68-72 °F) is needed.

For active *R. pipiens* and other species, rooms at 22-25 °C (72-78 °F) are optimal.

For tropical species and for use as an isolated insectarium, two rooms should be available at 26-30 °C (78-86 °F). The insectarium must be provided with a strong vent fan, high-capacity air intake, and a thermostatically regulated heater to prevent undue temperature fluctuations. Such venting is needed to minimize odors and to reduce the possible occurrence of insect allergies among personnel (see Chapter X, Section B.2).

c. Description of Ancillary Rooms

The breeding and general laboratory rooms [maintained between 20 and 22 °C (68-72 °F)] need equipment typical of a biological laboratory as well as extra shelf space for holding pans of embryos from fertilization to hatching. In addition to the usual equipment for autopsy and examination, this room needs a small refrigerator for the storage of carcasses until time of autopsy and disposal.

The equipment for the insectarium as described above, depends on the insect species under culture. An ample storage and general service room should be provided for the storage of cages, food, sundry supplies, and washing equipment as well as simple shop tools. It will need a refrigerator and a deep freeze. In a large unit the washing operation should have a room separate from the storage and general service functions. An automatic tunnel washer is advisable, but care must be exercised to test the toxicity of detergents used in such washers. Larvae may be particularly sensitive to detergent "buildups."

Equipment in the office and information control center will depend on the nature of the records and the information to be handled (see Chapter VIII). Adequate space for file cabinets and shelves for record books is mandatory. The room should be suitable for computer terminal installation. In addition, this room or an adjacent room should contain photographic equipment and space to store preserved animal specimens. Each of the "wet" rooms should be equipped with sinks that have heavy-duty industrial capacity disposals and with facilities that meet standards for animal waste disposal.

d. General Specifications

Other specifications for the amphibian quarters include hot and cold water, steam lines for cleaning, plentiful waterproof electrical outlets at 110 V, gas, and high-pressure air and vacuum lines. The floors must be designed to permit thorough cleaning, minimize slipping, and allow for good drainage. Lighting adequate to the function of each room must be provided as discussed below. The distribution of these facilities to the several rooms will depend on the geometry of the suite and the uses of the several rooms.

Each room associated with the amphibian quarters should be made as insectproof as possible because insects may inadvertently escape into the room when insect or amphibian enclosures are opened. Insect proofing should include special baffles and sealing materials around doors. The arrangement used for weatherproofing doors is suitable. Doors should be equipped with self-closing devices to ensure prompt and effective closure. The amphibian quarters should be provided with entrance ways with two sets of doors that form an entrance chamber—another aid in the control of escaped insects. Floor and ceiling moldings should be carefully inspected for possible routes of insect escape, as should the points of entrance of pipes, wiring, etc. Measures must be taken to control insects that have escaped, but insecticides must not be used as they are a danger to the food insects and the amphibians. Cleanliness is the best control, but it may be aided by the appropriate use of flypaper.

Floors, walls, and ceilings should be of water-resistant material to permit hosing or steam cleaning at regular intervals. One of the major contaminating arthropods in the insectarium and the amphibian quarters is the spider, which finds the environment highly compatible because of the available food. The only method to control spiders is cleanliness. Other contaminating insects include parasitic Hymenoptera (wasps) and Diptera (flies). These, too, can only be controlled by cleanliness and care to prevent their introduction from the outside. In one amphibian installation, bats, thriving on the insect population, have constituted a nuisance, even during the winter months.

2. Environmental Control

a. Water

The water supply is most critical to a successful amphibian colony. For aquatic forms, e.g., anuran larvae and the aquatic urodeles, the requirements for water quality are as critical as those for fish; for the terrestrial and semiterrestrial juvenile and adult forms, water remains an important component of the environment. Thus, though critical tests have not been published, we recommend that water standards for both larvae and adults be held within the limits prescribed for fish. Although we review here some of those standards that seem particularly applicable to amphibians, we recommend reference to fish standards (Committee on Standards, in press) and texts on the engineering and biological aspects of facilities for the long-range and large-scale maintenance of fish (Spotte, 1970; Clark and Clark, 1971; Bardach et al., 1972). McKee and Wolf (1963), American Public Health Association (1965), and Culp and Culp (1971) are valuable references on the evaluation of water quality.

Water quality is highly variable between geographic locations and is affected by its source, quantity, method of transport, type of food placed in it, and amount of waste released into it (Bennett, 1962). In particular, differences may be expected between water from subterranean and surface sources, the latter more commonly possessing deleterious characteristics. The quantity of water consumed depends on the size of the facility, the number and types of animal and the flushing efficiency of the animal container (see Sections C.2, 4, and 6). Since container designs have not been standardized, the quantity of water needed for proper care of amphibians must be determined for each facility. Water should be available in quantities greater than use expectation. Where adequate supplies of running water of high quality are not available, facilities for recirculating water are essential (Spotte, 1970; Clark and Clark, 1971; Cullum and Justus, 1973); the capacity of such systems for supplying large facilities, however, is limited by the economics of filtering, treating, and pumping used water.

In view of these qualitative and quantitative differences between water supplies, the quality and flow rates of water should be monitored and the water treated to ensure qualitative acceptability. In planning the water supply, care should be taken to ensure its quality with respect to:

- Alkalinity and hardness as $CaCO_3$
- Ammonia and other nitrogen compounds
- Carbon dioxide
- Chlorine
- Fluorides
- Heavy metals
- Microorganisms
- Oxygen
- pH
- Polychlorinated biphenyls (PCB) and other toxicants from plastics
- Toxicants

Municipal systems are a variable source of water. For example, in a city with several wells and surface sources, it is not uncommon to shift between these sources on a seasonal or even daily basis. Consequently, water chemistry should be monitored regularly. Acceptable procedures can be found in the publication prepared by the American Public Health Association (1965). An inviolate schedule for such analyses should be established, and a specific member of the staff should be assigned responsibility for this service. Remember, however, that such monitoring can never cover measurement of all possible changes; changes in water quality, especially in the northern states, may be pronounced at the time of winter freezing and spring thaw. The changes may involve organic contaminants not normally detected by routine water quality assays. It is extremely important to note that such sudden changes in water quality may be highly deleterious when mating procedures are being conducted. Thus, as described in Chapter VII, we recommend that artificial media formulated from distilled water be used in containers for mating, artificial insemination, and early development.

(1) *Alkalinity and Hardness as Calcium Carbonate* Total alkalinity (total ionic strength) and hardness should be maintained between 150 and 250 mg/liter, compared with a 60-120 mg/liter standard sometimes recommended for public water supplies. The addition of food to rearing containers of larval amphibians should adjust the hardness and alkalinity values in the water and provide the larvae with the necessary minerals. If alkalinity or hardness must be adjusted because of deficiencies that may occur in some public water supplies or in places such as some of the mountain states or because of excesses that clog valve systems, water treatment specialists should be consulted. Rather sophisticated equipment is available, but even then it will be necessary to maintain tight control over the system.

Well water may be high in iron and a variety of salts and deficient in oxygen.

(2) *Ammonia and Other Nitrogen Compounds* Concentrations of ammonia above 0.2 mg/liter as nitrogen are detrimental to fish and may also affect amphibians. Ammonium carbonate and ammonium hydroxide form in waters high in carbonates. At 4 mg/liter these compounds are toxic to fish and can cause stressful pH changes. Thus, in recirculating water systems, ammonia must be kept at a minimum, especially in hard water.

Nitrates and nitrites are produced by bacterial decomposition of organic materials. Values for these compounds should not exceed 0.3 mg/liter as nitrogen; in the presence of phosphorus and wide spectrum lights algae growth, which may clog water pipes, will be promoted. Although problems with these compounds occur most frequently in recirculating systems, they will also occur in those areas where public water supplies are high in nitrates. Water should be checked for nitrates before use as they may be harmful to both larval and adult amphibians.

In flow-through systems the frequency of flushing or the rate of flow must be adjusted to prevent the accumulation of waste products and the bloom of putrifying bacteria whose actions have a variety of deleterious consequences. If such adjustment is difficult, it may be necessary to use "conditioned water" systems in which nitrifying bacteria control ammonia levels and in which populations of putrifying bacteria are depressed. An excellent account of the dynamics of well-established conditioned water systems is available in Atz (1971). Poorly defined "control agents" that seem to regulate *R. pipiens* populations (Richards, 1958, 1962; Rose and Rose, 1965; Gromko *et al.*, 1973) may accumulate in noncirculating water systems and limit the density of the tadpoles that may be cultured.

If nitrogen compounds must be removed, water treatment specialists should be consulted. However, it will be difficult to control nitrogen compounds without affecting other water quality characteristics. The water supply system may need reconstruction to obtain desired qualities.

(3) *Carbon Dioxide* Carbon dioxide should not exceed 5 mg/liter. It is doubtful that fish, and possibly amphibian larvae, can survive long periods exposed to 12 mg/liter of carbon dioxide.

In a well-aerated, flow-through system carbon dioxide is unlikely to reach detrimental levels. Increases in carbon dioxide, however, may depress pH values below desirable levels [see Section B.2.a(9) for a further discussion of pH].

(4) *Chlorine* Both chlorinated and nonchlorinated water must be available. Close attention to chlorine levels is needed as chlorine in public water supplies often will exceed lethal tolerance limits for aquatic amphibians.

Aquatic Water supplied to aquatic, gill-breathing larvae and adults or to hibernating or skin-breathing adult amphibians must be free of chlorine. Although some larval amphibians can tolerate chlorine levels as high as 3.8 mg/liter over a period of time, growth and other physiological processes may be affected. Thus, concentrations should be well below this level. Activated charcoal filters, aeration or sodium thiosulfate easily remove chlorine. For small operations, holding the water in tanks with large surface areas, bubbling air through the water, or agitating the water will remove the chlorine in a few hours. For intermediate-size operations, sodium thiosulfate (6–8 mg/liter of water) can be added to large reservoirs or metered into continuous flow systems. However, control of the sodium thiosulfate level is essential as concentrations of 5 mg/liter are known to be toxic to some fish.

For large facilities, e.g., 20 gallons of water/h or greater, cylinders of activated charcoal can be installed directly in the water supply system. Most commercial water treatment services can supply, install, replace, and service the charcoal cylinders for a nominal fee.

Copper chloride at 9 mg/liter has been reported as toxic to fish. Thus, copper water lines should be avoided; if this is not possible, water flowing through them should be closely monitored (see below).

Terrestrial The presence of chlorine in water provided for lung-breathing amphibians will retard bacterial growth and is thus beneficial. Nonhibernating adult *R. pipiens* can tolerate levels between 4 and 6 mg/liter and levels as high as 12 mg/liter for short periods (Kaplan, 1962). At the Louisiana State University facility, *R. catesbeiana* also tolerate 4 mg/liter. Higher levels have not been tested.

Because chlorine may be lost between the chlorination plant and the point of water usage, it may be necessary to meter chlorine into the water supplied to adults to control the levels of microflora. Chlorine gas is extremely dangerous and should be handled only by trained personnel. A high-quality flow meter is essential for accurate metering of chlorine into the water supply, although such meters require regular maintenance to ensure accurate delivery.

Amphibians such as *R. catesbeiana*—which tolerate chlorine but are fed underwater such living foods as fish or earthworms that cannot tolerate chlorinated water—must be handled differently with regard to chlorinated water. Their water supply must be shifted from the chlorinated to the dechlorinated line at the time of feeding.

(5) *Fluorides* Concentrations of fluorides should not exceed 1.5 mg/liter (Kaplan *et al.,* 1964). In northern climates, concentrations in water supplies may slightly exceed this value.

(6) *Heavy Metals* Heavy metals such as zinc, copper, mercury, and lead may enter food or water systems from many sources and must be evaluated before amphibians are reared.

Zinc may be leached from galvanized pipes, copper from copper or brass pipes, etc. Copper is toxic to gill-breathing organisms (see above), and zinc is known to be toxic (Pickering and Vigor, 1965) and to accumulate to lethal levels when fish are exposed to $ZnCl_2$ (McKee and Wolf, 1963). Although the toxicity of these metals to amphibians has not been evaluated, the aquatic forms may be at risk; pipes in the water system should be made from black iron or high-density polyethylene or polypropylene or nylon [see also Section B.2.a(10)].

Well water may be high in iron. Upon aeration the iron normally precipitates and is thus nontoxic. However, large quantities of precipitated iron may clog sensitive water valves or enhance growth of iron bacteria that may, in turn, clog valves or deplete oxygen in the water.

Municipal water departments may add copper sulfate to water supplies to control algal growth, particularly in the fall and spring. Copper sulfate is an inhibitor of tadpole growth. It may be removed by adding versene (EDTA) at 50 mg/liter of water (Richards, 1958).

(7) *Microorganisms* Some aspects of the role of bacterial flora in water supplies are noted above. Fecal coliform densities should not exceed 2,000/100 ml and total coliform not exceed 20,000/100 ml.

(8) *Oxygen* Gill-breathing amphibians must have an adequate supply of oxygen to survive normally. Since the oxygen requirements for aquatic stages of amphibian species have been poorly documented, the oxygen requirements for fish should be followed. For warm water fish, oxygen levels should not fall below 5 mg/liter and for cold water fish, 8 mg/liter. This suggests that larval stages of amphibians from northern climates may have higher oxygen requirements than the same species from more temperate regions.

Though gills are replaced by lungs during metamorphosis, oxygen should still be maintained at the recommended levels to prevent other complications in water quality. Should the water become anaerobic, bacterial populations will increase, and the chance of disease outbreak increases. Also, ammonia levels may increase to toxic levels.

Well water, depending on its source, may be either deficient in oxygen or contain an excess of oxygen that leaves in a gaseous form as the water warms. In either case, the water should be stabilized by aeration before use.

(9) *pH* A pH value outside the range of 6.5–8.5 may be detrimental to amphibians that remain in water for extended periods, and there is evidence that tadpoles of some *Rana* species develop best if the pH is around 6.5. However, many amphibians have evolved in natural environments with lower or higher pH levels and may have other optimum levels. For wild-caught animals, especially eggs or larvae, pH levels similar to those of the environment in which the animals were collected should be maintained.

Changes in ammonia and carbon dioxide concentrations can cause changes in pH [see Sections B.2.a.(2) and (3)]. In cases of pH values being depressed by excess carbon dioxide, correction can be attained by the addition of calcium sulfate or sodium hydroxide. Conversely, the water in some public supply systems may reach pH values as high as 9.5-10.5 because of the source or method of treatment. These pH values stress amphibian larvae; this state can be reduced by the monitored addition of acetic acid, although caution should be exercised as "iron bacteria" may clog pipes at lower pH values.

(10) *Polychlorinated Biphenyls (PCB) and Other Toxicants from Plastics* The ease and favorable cost of plastic piping, containers, and instruments recommend their use in many applications in amphibian quarters. However, they are not without danger, especially when in contact with the water supply. Phenolic and acrylic plastics may contribute significant levels of polychlorinated biphenyls to the water. The toxicity of these substances has been well documented for living systems and should be avoided. Pliable plastics contain up to 40 percent by weight of plasticizers, some of which are volatile. Phthalate esters of various kinds may be leached into water from these plasticizers. These substances have known toxic effects and should be avoided (Napier, 1968; Jaeger and Rubin, 1973; Krauskopf, 1973). Some plastics incorporate fungicides. In the absence of tests evaluating the effects of these fungicides on amphibians, these plastics should be avoided.

Thus, we recommend that, where plastic piping is used to avoid copper contamination, those made of high-density polyethylene, polypropylene, or nylon be used. If plastic containers for embryos and larvae are used, these should be of rigid plastics with reduced plasticizer content.

(11) *Toxicants* Toxicants are too numerous to describe here in detail, but any source of water or food contains potentially toxic substances. Insecticides are most likely to occur in commercial food preparations and should be quantified and possibly removed (Stober and Payne, 1966). Amphibians normally are tolerant of the insecticide concentrations found in commercial feeds. However, the insecticides may accumulate if appropriate precautions are not taken and may be detrimental in breeding colonies where there is risk of pesticide accumulation in fat-rich ova. Cooke (1971) has reported the toxic levels of pesticides for *R. temporaria* and *Bufo* larvae. It is not yet possible to define toxic levels of these substances, particularly in the absence of information on their synergistic action.

Common table salt is widely used as a general treatment for diseased fish and amphibians. Care should be exercised in use of such salt treatments as amphibians, particularly juveniles and adults, rapidly absorb these potentially lethal salts. Tolerance limits for the many species are not known, and it is best to avoid the use of salts on a colony of amphibians without first obtaining tolerance limits for a few individuals.

b. Temperature

Temperature control is necessary at all stages of amphibian life cycles. However, since the optimal temperature for different species or the same species at different stages of development or geographical locations varies, sufficient flexibility should be built into the water systems to permit regulation over the temperature ranges described below.

Unless the volume and flow rate of water and the insulation of animal containers are sufficient to prevent the water temperature from reflecting the air temperatures, air conditioning will be required. It is for this reason that recommendations were made above (see Section B.1.c) for animal rooms held at temperatures appropriate to each species.

Although temperatures for optimum growth and differentiation have not been well established for most amphibians, investigations at the Louisiana State University amphibian facility indicate that specimens of the same species collected from geographical regions separated by only 4° lat. and 458 m (1500 ft) elevation have different growth responses at a given temperature. Below 21 °C (70 °F) larval and juvenile *R. catesbeiana* collected at 30° lat. and 9.15 m (30 ft) elevation (Baton Rouge, Louisiana) did not grow as rapidly as those collected at 34° lat. and 460 m (1500 ft) elevation (south central Arkansas). Adult *R. pipiens, R. catesbeiana*, and *R. clamitans* from their northern ranges are active and readily tolerate 22-24 °C (72-75 °F); from their southern ranges, temperatures as high as 30 °C (86 °F) are tolerated. Toads may also be kept at these higher temperatures. *R. sylvatica* should probably be kept at 20-22 °C (68-72 °F)—a range that is also appropriate for *B. orientalis*. *Xenopus* can be kept over a wide range but do best at approximately 22 °C (72 °F). Axolotls do best when held at temperatures in the neighborhood from 21 to 22 °C (70-72 °F).

c. Lighting

The animal rooms should be provided with lighting that has spectral emission similar to natural sunlight and be equipped with timing devices to control photoperiod. Such lights and equipment are available from several manufacturers. Many enclosures may be constructed of materials that pass light poorly, selectively, or not at all, and the geometry of racks bearing the enclosures frequently allows little ceiling light to enter the enclosures. Careful design is necessary to house efficiently the animals in the space available and also permit adequate lighting (see Section C.6.d).

Little is known of the effects of light on amphibians. Guyetant (1964) reports increased growth and reduced mortality of wild-caught *R. tempo-*

raria tadpoles in constant light, but tadpoles from induced reproduction had most rapid growth in total darkness for 1 mo. Mortality among *R. pipiens* albino tadpoles is greatly reduced if they are maintained in the dark (Smith-Gill *et al.*, 1972). Preliminary evidence also suggests that laboratory-reared and laboratory-bred amphibians attempt to adjust their physiological states when isolated from normal environmental cycles, including photoperiod (Hejmadi, 1970).

According to Bennett (1962) and Reid (1961) amphibians hibernating under ice are unlikely to be exposed to wide spectrum lighting, particularly if water depths are greater than 1 m (3.3 ft) or if a blanket of snow covers the ice-coated pond. However, direct observation reveals that northern frogs hibernating at 1–5 m (3–16 ft) in lakes do not burrow in the silt. Photographs have been taken even under ice with snowcover without the aid of additional light (Emery *et al.*, 1972). Such animals should be dimly lighted on a winter photoperiod even in hibernation (see Sections C.5 and 6.a). In less severe areas where frogs are not forced into deep lakes they hibernate or aestivate by burrowing in mud of swamps or pond bottoms. Still others, such as *R. sylvatica* pass the winter under forest floor litter. In such cases, hibernation quarters should be dark.

The onset of reproductive cycles for many animals is triggered by the length of photoperiod, and this may also be true for amphibians. Until the effects of photoperiod on amphibians are more fully investigated, amphibians maintained under laboratory conditions should be exposed to the photoperiod of the habitat from which they originate (Mahoney and Hutchison, 1969; Hutchison and Kohl, 1971).

C. ENCLOSURES

1. Embryos: Fertilization through Initiation of Feeding

Shallow enamel pans, glass trays or finger bowls, fiberglass trays, frames covered with muslin or nylon mesh (see Sections C.6.c and d), or plastic pans [but see Section B.2.a(10) regarding the danger of plastics] may be used to conduct artificial fertilization and to carry embryos until after feeding has started. Conditioned water or, preferably, reconstituted pond water [e.g., 10 percent Steinberg's solution as described in Chapter VI, Section B.1.a(1)] should be used. Since in controlled experiments and mating protocols, great care must be exercised to prevent mixing of embryos from different clutches, trays with sides two to three times higher than the depth of the medium are desirable to minimize the possibility of accidents by sloshing of media and embryos from one container to another when the trays are moved. Other aspects of embryo culture are discussed in Chapters VI and VII.

2. Larvae

Enclosures for larvae must be designed to permit optimal larval growth, minimize maintenance labor, maximize density for efficient use of laboratory space, permit positive identification of fertilization batches, minimize the possibility of mixing individuals from fertilization batches, and prevent vigorous tadpoles from leaping from the containers. For small operations when the labor is conducted by the principal investigator or a well-trained assistant, the enclosures used for the embryos may continue to be used until metamorphosis. The number of tadpoles per liter, however, must decrease as their size increases.

For intermediate and large operations a variety of choices is available, each with benefits and defects. These choices involve the selection of water supply systems, fabrication materials, and enclosure configurations. Since the choice of fabrication materials and enclosure configurations depends on the character of the water supply system, the latter should receive initial consideration. Once-through continuous flow, recirculated, or "balanced aquarium" conditioned water systems are possible. For each of these, water quality standards must be met as noted in Section B.2.a, where the need for monitoring water quality and the equipment associated with its handling are discussed.

Where high-quality water is plentiful and space limited, once-through continuous flow systems permit high animal density and minimum direct labor. The labor is needed to control the flow system, ensuring that water is maintained at high quality, pressure is maintained at a constant level, and feed and drip lines are free of stoppages. The disadvantages are that water quality or flow rate may change drastically at night or on weekends, which could result in heavy animal losses, and that, if the water flow is too rapid, the animals do not have the opportunity to "condition" their environment.

Where water is less plentiful, recirculation systems (see Section C.6.d) are useful (Justus and Cullum, 1971; Cullum and Justus, 1973). These have many of the advantages of once-through continuous flow systems and all their disadvantages. In addition to the labor enumerated above, however, filters and other water treatment devices must be carefully maintained. Although the animals may "condition" the water to some degree, the possibility of dangerous accumulation of wastes exists, and little is known about the effective removal of possible "population control" substances. Space is needed for the filtration and treatment devices, of which cooling equipment may be the most costly.

Where water is scarce and large numbers of larvae will not be cultured simultaneously, "balanced aquarium" conditioned water systems are use-

TABLE 7 Criteria for Construction Material for Enclosures

Material	Advantages	Disadvantages
Fiberglass	1,2,3,4,5	11,12,13,15,21
Glass	1,2,3,4,5,6,7	12,14,20
Plastic	1,2,3,4,5,6,7,9	11,21
Plywood	2,4,6,10	11,16,17,19
Metal	2,4,5	11,13,15,16,17,18
Concrete or stone	2,8	11,13,16,17,18,19,21,22,23

Advantages: 1, Inert; 2, permanent; 3, many forms and sizes possible; 4, portable; 5, needs little space; 6, inexpensive; 7, transparent; 8, can be "conditioned"; 9, can be readily modified; 10, readily made.
Disadvantages: 11, Opaque (or translucent); 12, fragile; 13, commercial fabrication needed for special units; 14, adaptation may require professional help; 15, relatively expensive; 16, forms or designs limited; 17, requires "inert" sealer that may be toxic; 18, may be toxic if sealer breaks; 19, requires space; 20, potentially hazardous; 21, potentially dangerous constituents; 22, too heavy for "stack racks"; 23, no portability.

ful (see discussion in Section B.2.a, and Committee on Standards, in press). They are the simplest to monitor and maintain, but in their simplicity require careful supervision by a professional biologist or experienced fancier and thus are less well suited to the research laboratory environment. Such systems fall in the "mimic of nature" category discussed above (Section A), and because "balanced" water tanks usually contain vegetation and invertebrate scavengers that may serve as intermediate hosts for parasites, animals raised under these conditions must be classified as wild-caught animals (see Chapter III, Sections B.2 and 3).

Enclosures suited to these several systems of water supply may be fabricated from several materials, each having advantages and disadvantages (Table 7). The choice among these materials is ultimately dependent on the water system and enclosure design most suited to the circumstances and objectives of the amphibian quarters. Several design possibilities for enclosures for larvae are described in Section C.6.

3. Juveniles

a. At Metamorphic Climax

One of the most critical stages in the life history of anurans is the period between emergence of the forelimbs and the complete loss of the tail. Mouth structures undergo extreme modifications during this period and lung breathing becomes mandatory. The animals drown if terrestrial areas are not available. They do not eat and their locomotion is inept. The considerations noted above (Section C.2) regarding water supply and materials for enclosures apply at this stage.

Several design possibilities for housing animals at this stage are given in Section C.6. Among the simplest are enclosures such as vegetable crispers. These should be floored with solid core neoprene mesh to prevent the animals from being caught in the surface tension at water–plastic interfaces (see Section C.6.a). The enclosures should contain some water and should be set on a raiser such that one end of the container is 20–30 mm (0.8–1.0 in.) higher than the other with the water forming a pool covering about one third of the container area. Although once-through continuous-flow devices are advisable to minimize bacterial buildup (van der Waaij, in press), such containers require careful engineering of overflow safety valves and constant care to prevent flooding and drowning the animals. Therefore, static water is recommended until economically feasible and safe devices are available. When static water is used, it should be changed three or four times per week or more frequently to keep bacterial counts below toxic levels [see Section B.2.a(7)]. The water used at this stage should be from the same supply used for tadpoles. Chlorinated water should be avoided until the animals have completed their adjustment to the terrestrial habitat.

b. Postmetamorphic

When ready to take food, the juveniles should immediately be moved to larger containers (see Section C.6). Consideration of water supply and materials for enclosures are as described above (Section C.2).

In nature, juveniles of most anuran species used in laboratories forage on land. Although active, they are not yet adept at capturing food; thus, the enclosure should maximize surface area, but the headroom inside the enclosure should be limited to facilitate capture of food. A tiered enclosure is most effective at this stage; it is compact and provides maximal terrestrial area in proportion to the aquatic area.

These animals are not good swimmers, in fact they may drown easily, and are highly sensitive to water contamination. A once-through, continuous flow system or recirculating system with a water depth just sufficient to flood the floor of the enclosure is most appropriate until the animals have gained size and strength. Special attention must be given to the design of the drain to prevent clogging and escape of the animals (see Section C.6). This drain must also allow easy adjustment of water depth; deeper water is needed during feeding periods for amphibians, such as *R. catesbeiana,* that feed in water.

The "wet" floor of the enclosure should be covered with unglazed pottery bits to allow the juveniles to get out of the water and to reduce the loss in the water of young crickets and sowbugs used for food. These bits should be too small for the juveniles to hide beneath. On top of this layer should be placed a few large shards to provide sanctuary for the animals and to serve as islands for their escape should the water depth increase unexpectedly. The water at this stage should be the same as that supplied to the adults.

As soon as the animals have adjusted to terrestrial behavior, their enclosures should be modified in accordance with the design configurations for adult animals. If this is not done, growth, at least of *R. pipiens*, is inhibited; the animals lose their adaptability, and later transfer between enclosures becomes difficult.

4. Adult Enclosures (Evaluation Criteria)

Housing for adults must meet variable requirements adapted to the differing needs of different species and to the methods chosen to meet those needs. Water and material requirements are as previously described in Section C.2. Food preferences and methods of delivery, differing requirements for access to water and light, and differing needs for sanctuary are major concerns in selecting a design for adult enclosures. The following are important criteria that must be met in housing design: water flow patterns, ease of cleaning, accessibility for servicing and for identifying and handling individual animals, security regarding both amphibian and "food" escape, safety regarding potential injury to either the amphibians or personnel, efficient space utilization, modular design to permit flexibility of colony size, and acquisition and maintenance cost reduction. Evaluations of facilities for amphibians should include consideration of each of these animal- and facility-oriented desiderata (see Section C.6).

Ideally, it would be most desirable if one, or even a few, designs of prefabricated adult enclosures were available to take advantage of the improved quality and cost savings possible through mass production of adult housing systems. Unfortunately, such prefabricated systems are not yet available on the market, although designs for such systems have been prepared on the basis of current experience with large colonies of amphibians.

Until standardized low-cost adult housing systems are available, each facility maintaining adult amphibians will certainly utilize the most economical components locally available. At a minimum such "jerry-built" units must meet the animal-oriented requirements noted above, even at the expense of the facility-oriented desiderata. Section C.6 describes several adult housing systems currently in use. Undoubtedly, these will change as new information becomes available and as new techniques, particularly in the area of pelletized food delivery, are developed.

5. Hibernation Quarters

The design of quarters for hibernating animals depends on the species and on the number of animals to be maintained in hibernation (see Chapter VI, Section B.1). The major objectives of such quarters include the following:

- maintaining the desired temperature;
- allowing adjustment of the temperature through the hibernation period;
- permitting submersion in water in a manner appropriate to the normal habitat of the animal;
- maintaining adequate oxygen levels;
- providing adequate, low intensity, short photoperiod lighting;
- minimizing agitation of the animals and thus conserving their energy stores; and
- allowing removal of wastes and dead animals without causing stress to the living animals.

For animals such as *R. sylvatica* that hibernate under forest litter, appropriate containers can be provided with leaves and forest litter. This litter should be sterilized to destroy metazoan parasites prior to introduction into the animal enclosures. At present, optimal temperatures and moisture levels for North American animals requiring these conditions are unknown; based on the experience of the University of Hiroshima Laboratory for Amphibian Biology, however, 7-10 °C (45-50 °F) and frequent sprinkling with water seem to be adequate.

For the maintenance of small numbers of animals of the species that hibernate in deep water, it is perhaps most economical to use disposable plastic containers with water to a depth of 100-125 mm (4-5 in.). Water should cover the animals as it does in nature. This water must be chlorine free and should be changed approximately once a week or three times in 2 weeks using prechilled water. The water-changing schedule should be designed to reduce agitation of the animals and yet maintain bacterial counts below toxic levels [see Section B.2.a(7)]. Do not place these animals in refrigerators containing volatile materials.

Two options are now available for the hibernacula of larger numbers of animals. One involves large fiberglass tubs equipped with a cooling-circulating device. This can accurately regulate the water temperature and coldrooms need not be used. Such a system must also include aeration and filtering devices that remove wastes yet allow gentle water flow that does not agitate the animals. Several thousands of animals may be maintained in such a container (see Chapter IV, Section B).

A second method utilizes large, plastic garbage containers placed in coldrooms as hibernation enclosures. These are more fully described in Section C.6.a.

Regardless of the equipment selected, it is most important that amphibians that hibernate under ice be in water deep enough to cover them, that temperatures be maintained between 2 and 3 °C (36-37 °F) and *not above 4 °C (39 °F)*, that the water be well aerated, that low-intensity light levels be maintained for 8-10 h, and that agitation of the animals be held to a minimum.

6. Enclosure Designs

In the absence of definitive standards for amphibian enclosures, this section describes several aspects of the housing systems now in use. It is hoped that it will meet the demand for specific designs to guide the planning of those developing facilities for amphibians. Sufficient guidelines are given above and in Chapter VI for the user of amphibians who handles only a few animals for short periods. Such a facility is not described here; the reader is referred to Schmidt and Hudson (1969). The descriptions presented are not exhaustive (see also Boterenbrood, 1966; Frazer, 1966) and the reader must evaluate them with respect to the animal- and facility-oriented desiderata listed in Section C.4. The serious planner should visit one or more of these facilities before investing heavily in equipment for the care of laboratory amphibians.

a. The Amphibian Facility of The University of Michigan

The housing and management system used in this facility is thoroughly described in Nace (1968); although improvements have changed some specific operations, that document remains essentially current. It is a flow-through system designed to house both anurans and urodeles that number in the thousands. Major emphasis is on *R. pipiens,* but significant colonies of *X. laevis* and *B. orientalis* are also under development. Small test colonies of approximately 10 other anuran and urodele species are also maintained. The colony is managed and data manipulated with the assistance of computer-based techniques (Nace *et al.,* 1973).

Figure 14 illustrates the enclosure used to house larvae of all anuran species from the initiation of feeding until the emergence of forelimbs. Figure 15 shows a portion of the five-tiered rack that carries these enclosures. One rack measuring 0.46 × 2.44 × 2.44 m (1.5 × 8 × 8 ft) carries 130 enclosures. Each enclosure houses 50-75 larvae initially, which are thinned to 15 per enclosure by the time of metamorphosis. Thus each rack carries from 2,000 to 9,750 larvae, depending on their developmental stage, or 166–800 larvae per square foot of floor space occupied by the rack. This configuration allows precise control of water flow, easy cleaning by frequent flushing or by enclosure replacement, isolation of small or large groups of larvae of known identification, ready access for servicing, safety, efficiency of space utilization, modest installation costs, long-term service, and low maintenance costs. The growth characteristics of larvae in these enclosures closely resemble those of larvae in nature.

A rack of enclosures for juveniles and adults is shown in Figure 16. The enclosures consist of transparent plastic mouse containers inserted into similar opaque containers. Each measures 0.19 m (7.5 in.) deep, 0.24 m (9.5 in.) wide, and 0.45 m (18 in.) long, but in combination their depth is 0.25 m (10 in.). Each combined enclosure may contain up to 30 or 40 juvenile *R. pipiens*. A rack measuring 1.13 × 2.44 × 2.44 m (3.7 × 8 × 8 ft) carries 76 of these enclosures for a capacity of approximately 30 juvenile frogs per square foot of floor space. Similar racks carry containers that measure 0.20 m (8 in.) deep, 0.39 m (15.5 in.) wide, and 0.50 m (20 in.) long, but 0.35 m (14 in.) deep when combined. These are used to house from 20 to 100 adults per combined enclosure. The smaller number of occupants is used when maximum growth is desired; the larger number when kidney tumors are being induced. A rack carries 36 such enclosures for a floor density of 24–120 adult frogs per 0.09 sq m (1 sq ft). An illustration of a disassembled enclosure is shown as Figure 4 in Nace (1968).

Each combined enclosure contains water in the opaque portion to a depth appropriate to the behavior of the amphibian species it contains. An opening in the floor of the transparent component allows frogs to move between the aquatic environment and the terrestrial environment of the transparent component. The latter contains a high, dry shelf, neoprene mesh on all floor surfaces, and several shards of unglazed pottery. One of several insect-proof lid designs is shown. Each enclosure is placed on sliding-arm runners to facilitate access and maintenance of the heavy, water-filled containers. For ready access, water control valves are placed at the front immediately above each enclosure, but a tube guides the inflowing water into the back of the enclosure. Figure 17 illustrates a drain tube— either a trombone-slide device or a wire reinforced hose—which drains each enclosure toward a trough running the length of the rack at each tier and receiving drainage from enclosures on each side of the rack. Access to the animals is possible either through the opening in the lid or by separation of the transparent from the opaque component of the enclosure. Food is introduced in appropriate containers placed on the floor of the upper component.

Xenopus and other aquatic amphibians may be held in these enclosures by removing the transparent component and placing the lid on the opaque

66

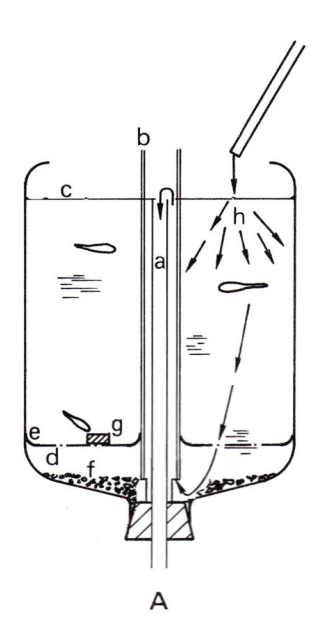

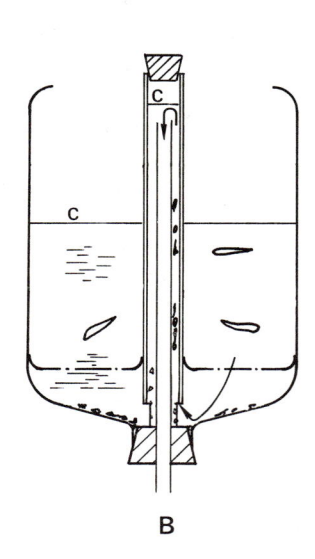

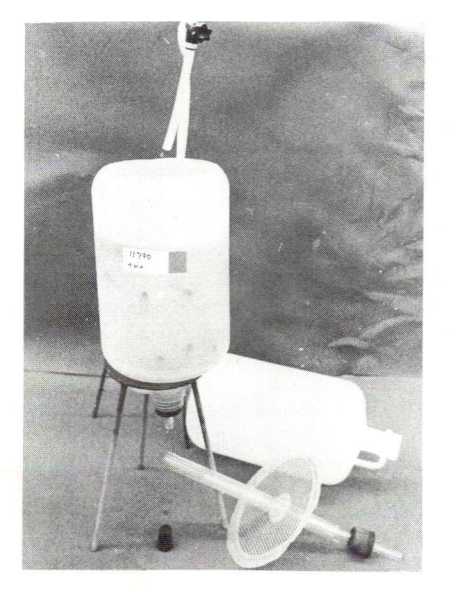

A
B
C

component. Removal of wastes by flushing may be supplemented by the use of water vacuum tubes.

Figure 18 diagrams a hibernation enclosure. These enclosures are placed in a "4 °C" coldroom in which temperatures oscillate between 0 and 4 °C (32-39 °F). The 20-gal capacity of the enclosure minimizes the temperature fluctuation experienced by the frogs and usually is stable between 2 and 3 °C (36-37 °F). Frogs received from the wild during the winter have usually been exposed to "room temperature" for at least several days. They are washed free of packing material and placed in the hibernation enclosure in water at room temperature. The capacity of the enclosure ensures that the water temperature does not drop to hibernation temperature faster than the animals can adjust to it. The pump and filter device ensure clear, aerated water circulated continuously by a gentle flow that does not agitate the hibernating frogs. Every 7-10 days, 2-3 gal of water are drained from the bottom to remove collected debris. This water is replaced with fresh, prechilled water.

Frogs remain at the bottom of the enclosure, particularly when the light is on. When the light is off, they sometimes swim to the surface. As many as 100 wild-caught *R. pipiens* females that retained usable eggs have been held in hibernation in this type of enclosure from October to July.

FIGURE 14 Diagram of the larvae enclosure. A. Flow-through configuration: (a) a 10-mm plastic stand-pipe fixes water level. It is set in a rubber stopper readily retained when firmly inserted into the neck of the enclosure; (b) a 15-mm-diameter stiff, open plastic tube with legs assures mixing of incoming water; (c) water level; (d) a barrier screen of stainless steel (2 mesh sizes for different stages) retains tadpoles but passes debris; (e) plastic cuffs on the screen assure contact with (b) and with the sides of the enclosure to prevent escape of tadpoles into the neck of the enclosure; (f) debris collected in the neck of the enclosure. The bottle should be one with a steep slope in this region to assure that debris is cleanly removed by flushing; (g) a food pellet; (h) the pattern of water flow is indicated by the arrows. B. Flush configuration: A stopper is placed in the open tube to change the drain system from a flow-through, water-mixing to a flush device. The outer tube is pumped up and down several times to initiate siphon action. Debris is siphoned from the neck of the enclosure, an action aided by twisting the bottle to loosen the debris. On completion of siphoning, water may be poured in to return the enclosure to overflow level or it may return to this level more slowly by drip addition. C. Assembled and disassembled: Each enclosure, fabricated of round 1-gal plastic bottles with the bottoms removed, contains approximately 3 liters of water, and water is dripped into the enclosure at a rate of approximately 3 vol per day. Flushing removes and replaces approximately 0.5 vol and is conducted once or twice a day depending upon the developmental stage of the larvae. Thus, larvae are in an effective vol of approximately 0.2-1.0 liters per larvae per day depending upon their stage.

FIGURE 15 A portion of a rack of tadpole enclosures.

FIGURE 16 A portion of a rack of enclosures for juvenile and adult frogs.

b. The *R. catesbeiana* Facility at Louisiana State University

This housing and management system is a flow-through system designed to house *R. catesbeiana* for a program to test management and husbandry concepts. The colony is comprised of laboratory-reared and wild-caught conditioned animals.

Rearing enclosures, fabricated from 1.22 m (4 ft) diameter round or oval galvanized cattle watering tanks coated with epoxy paint, can be used to house up to 100 juvenile frogs with a 76.2-mm (3 in.) snout-vent length. Coating is necessary to prevent possible toxification by zinc. The enclosures are inclined at an angle so that three fourths of the tank floor is covered with water with a depth of from 10 to 40 mm at the deepest point. A drain line is placed in the side about 40 mm from the bottom. A plastic tubular net is inserted in the drain to prevent food (fish) from escaping. If shallower water is desired, a drain can be cut in the bottom and an overflow tube (vertical) installed at the proper height. Such a drain also aids in flushing the tank periodically.

As the frogs grow, higher sides are required to prevent escape. Thin aluminum or plastic sheeting is attached to the sides to increase the height of the walls to 1 m (3 ft). The extension does not completely encircle the tank as it would be impossible to lean over the tank and work with the frogs. For this purpose a 0.6-m (2 ft) section is left open on one side.

Small enclosures used to house experimental groups of frogs are illu-

A. Expanded   B. Assembled

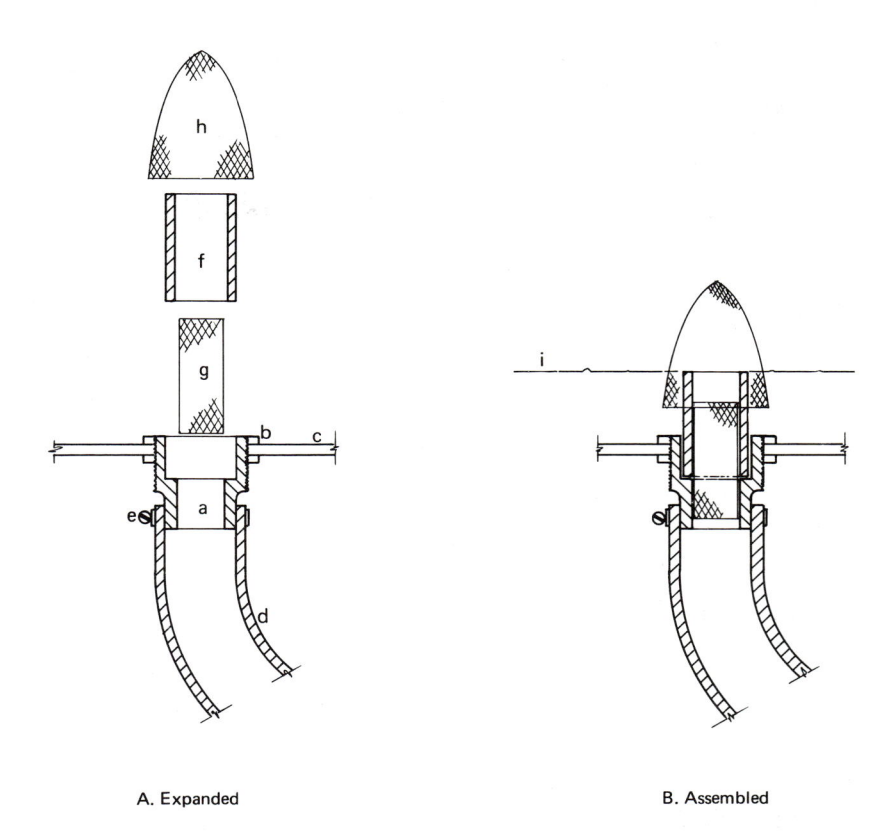

FIGURE 17 Drainage device used in juvenile and adult enclosures. (a) Nylon drain 20 mm in diameter; (b) set nuts and gaskets holding the drain in the floor; (c) of the enclosure; (d) drain hose held in place by hose clamp (e); (f) variable length, stiff garden hose which snuggly fits the offset in the nylon drain and whose length determines the depth of the water; (g) a plastic mesh sleeve which snuggly fits the inside of the hose (f) and into the narrow portion of the drain. When the hose (f) is lifted to rapidly drain the enclosure, this sleeve forms a barrier preventing juvenile frogs, carried in the water flow, from passing through the drain or, if large enough, from clogging its opening; (h) a plastic mesh sleeve formed into a cap by heat sealing. It is larger in diameter than the hose (f) and long enough not to be dislodged. It prevents juveniles from escaping through the hose. This cap need not be used in enclosures for frogs too large to pass through or become lodged in the hose; (i) water level. When only a film of water is desired, hose (f) is removed and cap (h) is used in combination with a longer inside sleeve (g). The depth of the water is then no greater than the thickness of the set nut (b).

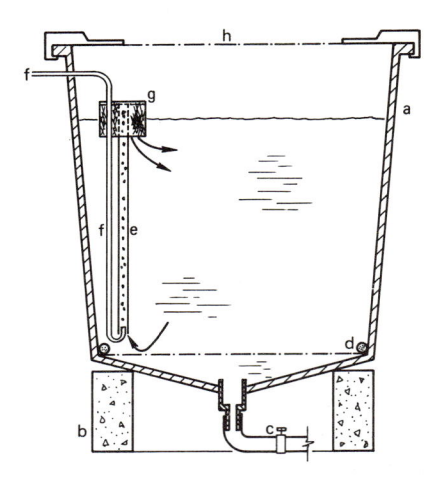

FIGURE 18 Hibernation enclosure. (a) a 20-gal plastic "garbage" container; (b) a ring of cement blocks raises the enclosure and permits the weight of the water to form the enclosure bottom into a shallow funnel shape; (c) nylon drain, hose, and clamp; (d) a heavy plastic screen weighted down by a sealed ring of plastic tubing filled with "shot" keeps the frogs from occluding the drain and forms a reservoir for debris; (e) a 15-mm-diameter stiff plastic tube extends to a point just above the screen; (f) a 10-mm plastic tube from a compressed air line bubbles air into the larger tube (e). The combination of (e) and (f) forms an air-lift pump which aerates the water while lifting it to the surface; (g) a plastic container filled with fiberglass serves as a filter. The bottom of this filter is perforated. It is attached to the side of the enclosure above the water-line to hold the filter and pump in place. Tube (e) passes through the filter and the aerated water lifted through it is spilled over the fiberglass and is filtered as it returns to the enclosure through the bottom of the filter; (h) the plastic lid of the enclosure contains a suitable screen to admit light whose duration is regulated by a timing device.

strated in Figure 19. These enclosures house up to 25 frogs with a 76.2-mm (3 in.) snout-vent length or 50 metamorphosing larvae or juveniles. Two types are used. One measures 0.28 × 0.33 × 0.14 m (11 × 13 × 5.5 in.), the other 0.45 × 0.73 × 0.20 m (18 × 29 × 8 in.). By tilting these containers, the elevated section provides a terrestrial environment. These enclosures require covers to prevent the escape of frogs but need not be insectproof as only fish are used in the diet of these animals. It is important, however, that the cover material not cause injury to the frogs. Plastic netting is better than metal screen or hardware cloth; metal screening may be used for strength if it is faced with neoprene mat.

If a shelter is provided in these enclosures, the frogs will move under it, making escape less likely during feeding or cleaning operations. Restaurant supply houses have a wide selection of plastic and fiberglass containers, some of which have sliding tops, that can serve in place of the enclosures shown in Figure 19.

The use of a ribbed, black, rubber floor padding in the dry portion of the floor of enclosures eliminates cuts and skin abrasions and facilitates frog movement without slipping. No flooring material has been found that is easy to clean and maintain, but use of some type of padding is advisable. Plastic or rubber netting is unsatisfactory as shed skin and dead fish get caught, making cleaning difficult.

c. Southern Frog Company (J. M. Priddy, Dumas, Arkansas)

The husbandry facility for *R. catesbeiana* at the Southern Frog Company uses a flow-through system that is a scale-up of the system used at the *R. catesbeiana* facility at Louisiana State University and follows designs originally developed by Stearns (1973). It is designed to produce commercially significant numbers of wild-caught conditioned and laboratory-reared animals (Priddy and Culley, 1971).

In place of cattle watering tanks, circular enclosures 6.1 m (20 ft) in diameter with 0.75–0.90-m (30–36 in.) walls of concrete block are used.

FIGURE 19 Small enclosures for test animals.

The flooring is concrete and slopes to one side with a drain line. Since lime leaches from concrete and causes skin erosion, a concrete sealer is painted on all inside surfaces and then covered with lead-free or other heavy-metal-free epoxy paint or swimming pool paint. These enclosures house up to 5,000 newly metamorphosed frogs or 1,500 with a snout-vent length of 76.2–127 mm (3–5 in.) for a floor density of 2–16 frogs per 0.09 sq m (1 sq ft). When animals in such numbers are placed in a single enclosure, sets of two concrete blocks placed in V patterns at several locations in the container are useful to minimize animal pileups that may occur when the animals are alarmed.

R. catesbeiana larvae are reared through metamorphosis in 4 months from fertilization at an average temperature of 28 $^{\circ}$C (82 $^{\circ}$F) using these same enclosures. One or more frames measuring 0.31 × 0.31 × 0.92 m (1 × 1 × 3 ft) are covered with nylon mesh and placed on 0.15-m (6 in.) legs in these 6.1-m (20 ft) diameter enclosures that are flooded to 0.31 m (12 in.) with continuously flowing water. The 200 larvae placed in each frame are fed a finely powdered minnow meal that floats. To prevent this feed from washing out of the nylon enclosures, it is placed inside floating frames made from styrofoam coolers with their bottoms removed. The tadpole swim under these frames and to the surface to ingest the feed. Food is given twice a day at the same time each day to condition the tadpoles to maximum feeding. *R. pipiens* larvae as well as other larvae may be reared by this technique.

d. The Aquatic Animal Facility of Arizona State University

The stainless steel housing system for anurans and urodeles of the Aquatic Animal Facility of Arizona State University utilizes recirculating water filtered through dacron and passed under germicidal lamps. This design requires a minimum of maintenance, provides control of temperature and light, and has been successfully tested as a holding facility for axolotls, frogs, toads, fish, crayfish, and turtles (Justus and Cullum, 1971; Cullum and Justus, 1973).

The volume of water in this system is 500 gal recirculated at the rate of 850 gal/h and exchanged by the removal of 11 gal/h and the addition of domestic water at 11 gal/h. Bacterial plate counts range from 600 in water draining from adult frog holding tanks to 15 in water after ultraviolet lamp treatment.

The design of the water circulation and processing system may be found in Cullum and Justus (1973). The following expands on their description of certain of the animal enclosures.

Figure 20 shows a breeding unit that measures 0.35 × 0.20 × 1.20 m

(14 × 8 × 48 in.). Newly fertilized eggs are placed in 0.23 m (9 in.) square stainless steel frames with fine muslin bottoms. These are set in the breeding tank until the tadpoles have reached swimming stages. They are then released into the tank that can hold 6,000 tadpoles up to 20 mm (0.78 in.) in length for a density of 1,250 young tadpoles per 0.09 sq m (1 sq ft). Members of different clutches are not separated and only one age group is carried per cycle.

Tadpoles are then moved to a rearing tank until metamorphosis is completed. This tank (Figure 21) measures 1.5 × 3.1 m (5 × 10 ft) and is 0.20 m (8 in.) deep. It can hold 8,000 tadpoles for a density of 160 larvae per 0.09 sq m (1 sq ft).

Upon metamorphosis, the frogs are moved to adult holding tanks. These measure 0.40 × 6.1 (1.3 × 20 ft) × 0.20 m (8 in.) and contain an aquatic area and feeding trays (Figure 22) that may be moved to allow the use of dividers to separate frogs of different categories. On the trays the frogs are fed mealworms and other crawling insects or crickets with their hind legs removed. The capacity of each adult holding tank is 80 frogs. Since these tanks are stacked in two tiers, the capacity is six frogs per 0.09 sq m (1 sq ft) of floor space.

The holding tank may be adapted to the housing of *A. mexicanum*. This system is also useful for other aquatic amphibians such as *Xenopus*. Perforated stainless steel baskets inserted in these tanks serve as enclosures for the aquatic animals.

FIGURE 20 A breeding unit for eggs and young larvae. Water is received in the upper compartment, passes to the larvae enclosure below, and is returned to the recirculating system by the components on the bottom level. Note the positioning of lights.

FIGURE 21 Interior of a rearing unit. The top is removed from the drain component. Water enters through the filter at the bottom of this component and exits via the stand-pipe. Terrestrial areas for metamorphosing frogs are shown adjacent to the drain component, the top of which also serves as a terrestrial area.

FIGURE 22 Interior of an adult holding tank. A feeding tray with access ramps extending both upstream and downstream is shown.

VI Amphibian Management and Laboratory Care

A. GENERAL COMMENTS

Care of amphibians is necessarily complex because of the different environmental requirements at different stages in their life cycles. In their premetamorphic or larval stages, they live entirely in an aquatic environment; upon metamorphosis they become partially or wholly terrestrial, with the exception of the newt which undergoes a second change and returns to water. Thus, any institution developing or maintaining laboratory colonies of amphibians must design housing facilities to provide for this dual requirement of aquatic and terrestrial environments.

Aquatic facilities require more care than terrestrial environments (see Chapter V). Sanitation and waste removal are crucial as the materials often are soluble and may directly affect exposed amphibians. Thus, large volumes of water for flushing are required or, if the water is to be recycled, treatment facilities must be installed and maintained. Water quality must be rigidly controlled as sudden physical or chemical changes may severely stress the animals. It is often necessary to develop special feed formulations since conventional feed may rapidly deteriorate and contaminate the environment when added to the water.

The period of transformation from an aquatic to a terrestrial mode of living requires both aquatic and terrestrial environments. Some species continue to feed in the aquatic environment; others become terrestrial in feeding habits; and still others may feed in either environment. Nutritional habits and requirements change as the animal moves from an aquatic to a terrestrial environment. Different methods of disease control must be developed depending on the type of environment. Housing, surgical methods, humane disposal, shipping and receiving, quarantining, and protection from injury must all be handled differently for animals in the two environments.

Finally, extreme care must be exercised in providing optimal environmental conditions during two delicate stages in amphibian life cycles, i.e., during embryonic development and metamorphosis. Rapid physiological changes occur at both periods and sudden environmental changes may prove detrimental or lethal.

Even though many problems in amphibian culture still exist, sufficient progress has been made to warrant description of culture techniques for several species (see also Boterenbrood, 1966; Frazer, 1966). Procedures for *R. pipiens* and *R. catesbeiana* will be described as prototypes. Where known variations exist between these and other species, these will be noted. These techniques do work as attested to by current success in maintaining colonies of amphibians. Nevertheless, there is much room for improvement and existing techniques will be improved and new techniques will be developed as our knowledge of the optimum conditions for rearing amphibians increases.

Most of the methods described here are designed for facilities maintaining less than 10,000 animals. Although more animals could be maintained with these methods, commercial dealers cannot economically afford to do so at present. However, incorporation of some of the husbandry principles could improve commercial stocks at little extra cost, and attention to the objective of stock improvement would certainly result in the development of improved techniques. A better quality animal commands a higher price. Once these animals and prices become established, commercial dealers should improve their facilities to include larger numbers of high-quality amphibians.

B. ANURANS

1. Ranidae

a. *R. pipiens*

    (1) *Premetamorphosis*

        (a) Enclosures   See Chapter V, Sections C.1,2,3, and 6.

        (b) Environmental Controls   See Chapter V, Section B.

        (c) General Care   Until feeding stages are reached, embryos are held at low densities in shallow pans. To provide a maximum surface-volume ratio, the medium should not exceed a depth of 15-20 mm (0.6-0.75 in.). Dead embryos should be removed regularly to avoid contamination, and the medium changed at least every third day or more frequently if it becomes turbid. Before hatching, 200–300 eggs/liter can develop successfully.

After hatching and when feeding begins, larvae should be thinned to 50 or less/liter; overcrowding will stunt growth (Richards, 1958, 1962; Rose and Rose, 1965; Gromko *et al.*, 1973). As the tadpoles grow, they should be continually thinned or given more medium so that near metamorphosis there are only four to six tadpoles/liter. The embryos and tadpoles should be housed in a container that provides a large surface area for gas exchange. If the volume-to-surface-area ratio is not adequate, the use of an air line to oxygenate the water is advisable.

During this period the use of artificial medium is recommended [for example, a 10 percent concentration of Steinberg's or Barth's modifications of Holtfreter's (Holtfreter, 1931; Rugh, 1965) medium or De Boers solution (see Chapter VII, Section A.10)]. The formulations of Steinberg's and Barth's media may be found in Johnson and Volpe (1973).

At this and all other stages, great care must be exercised for the cleanliness of instruments used in manipulating the embryos and larvae. Contamination with formalin, etc., must be rigorously avoided. Indeed, it is good practice to reserve sets of instruments for each clutch of animals.

For larger numbers of later larval stages, the tadpoles may be housed in any of several types of enclosures (see Chapter V, Sections C.2 and 6). Preference should be given to those types requiring minimum daily attention and minimal handling of the stock in order to reduce stress to the animals and the chances for error in labeling or in mixing tadpoles from different clutches.

(d) Food Supply Ranid larvae should be fed when strands of fecal material appear in the water some days after they hatch from their jelly. Young larvae tend to be vegetarians and the larger tadpoles are omnivorous. Because larvae may be raised on a variety of foodstuffs (Hamburger, 1960; Rugh, 1965; Di Berardino, 1967), the choice should be determined by the type of tadpole enclosure. Without special care, the food will float to the surface and produce a scum that inhibits gas exchange (see Chapter V, Section C.6.c). If the animals are housed in a water bottle system (see Chapter V, Section C.6.a), food must not be of such fine grain as to settle in the dividing screen or be lost as a result of the continual water flow. The food must not be allowed to disintegrate and decay in the medium—a problem that is particularly applicable to static water enclosures. For colonies with laboratory-reared or laboratory-bred animals, food of uniform quality must be available throughout the year.

Spinach must be avoided because it causes kidney stones in a number of amphibian species (Berns, 1965). Romaine or escarole lettuces have been found most suitable though their nutritional adequacy remains uncertain. Raw lettuce will not be eaten, but it may be readily softened in a pressure cooker and may be stored in a frozen state and dispensed to the animal containers with ease. Some caution in feeding is required since tadpoles are killed if their containers are overloaded with wilted lettuce. Since large tadpoles eat enormous quantities, it may be necessary to supply the containers with lettuce twice a day. Even in flowing-water bottle systems uneaten lettuce must be removed to prevent fouling of the water. Protein supplement is provided by feeding cubes of raw or boiled liver two or three times a week.

Tadpoles will also eat pulverized rabbit chow, "dry" dogfood, or a variety of other commercially available animal feeds. They will also survive on a variety of other diets, including powdered or hard-boiled egg yolk, raw liver, or liverwurst. These diets, however, are generally inappropriate because the foodstuffs rapidly disintegrate and decay in water. Thus not only is the water quality reduced but the material may be caught in the dividing screen if the flowing-water bottle system is used.

Tadpole feeds that incorporate commercial feeds and binders to prevent disintegration have been developed. These feeds avoid the difficulties encountered with the feeds described above. Such a feed was described by Hirschfeld, Richards, and Nace (1970). It is prepared by adding 250 g of pulverized chow, 20 g granular agar, and 14 g unflavored Knox gelatin/liter of water; the mixture is brought to about 100 °C (212 °F), after which it is allowed to solidify in flat pans. It can then be sliced for use or stored at -20 °C (-4 °F). It may be stored indefinitely at this temperature or for up to 14 days at refrigerator temperatures but should not be held at room temperature. Neither agar nor gelatin alone is satisfactory as a matrix. Rabbit chow incorporated in such binders has now been used for some years. It has good holding qualities in the flowing-water bottle system at the temperatures used for raising *R. pipiens*.

Experience has indicated that recently hatched tadpoles should first be fed wilted lettuce in the "embryo enclosures" for several days; then they should receive both lettuce and the agar-gelatin diets for several additional days, after which the agar-gelatin diet alone is sufficient. This diet does not cloud the water and requires minimal attention. A full 1- to 2-day ration can be administered at one time in that it is not excessive for a single feeding, and no debris remains for manual removal. [See Section B.1.b for a description of a similar preparation used for *R. catesbeiana* (Culley and Meyers, 1972).]

Various diets must be tested on the larvae grown in each facility. Variations may exist between animals from different original sources. Furthermore, postmetamorphic survival and maturation efficiency are greatly improved by adequate larval nutrition.

(2) *Postmetamorphosis*

(a) Enclosures   See Chapter V, Sections C.3,4,5, and 6.

(b) Environmental Controls   See Chapter V, Section B.

(c) Feeding Juveniles   The most successful diet for juvenile *R. pipiens* has been sowbugs or crickets, selected in sizes appropriate to the animals to be fed. (See below for information on handling these arthropods.)

Another suitable diet is mosquitos. Kawamura of the Hiroshima Laboratory for Amphibian Biology has routinely introduced mosquito tumblers into the containers with metamorphosed ranids. Upon eclosion, they are readily eaten by even the smallest froglets. The mosquito is an excellent nutrient source and can be readily retained within the containers. At the University of Michigan Amphibian Facility *Culex pipiens,* a bird mosquito that is easily raised according to standard procedures (Keppler *et al.,* 1965) with Japanese quail as blood donor, has been used successfully.

Other dietary regimens at this stage must be viewed with caution. Force-feeding of small, recently metamorphosed frogs has been found to be nutritionally inadequate and time-consuming. Live insects should be provided. Mealworms are frequently recommended, but separation of large numbers of small mealworms from their culture medium is onerous and they quickly die when placed in a moist environment. Furthermore, mealworms have been reported as nutritionally inadequate for *R. pipiens* (Cairns *et al.,* 1967). If mealworms are used, the frogs must be transferred to dry containers during the feeding period. This handling is deleterious and time-consuming. The low biomass of *Drosophila,* their inability to survive near water, and the difficulty in containing them does not support the recommendation they frequently receive as frog food.

(d) Management of Active Adults   The laboratory management and care of active adult *R. pipiens* require both aquatic and terrestrial environments as implied by the description of enclosures (see Chapter V, Sections C.4 and 6). All animals should initially be treated as described in Chapter IV, Section C.2, pertinent information should be recorded (Chapter VIII), and special attention should be given to the possibility of disease (Chapter IX). Those animals to be used within a few days for any objective other than egg production may be held for this short period in a box or covered aquarium at room temperature. The enclosure should contain enough water to permit the animals to submerge completely. It also should be provided with a table or shelf about 10 mm (0.5 in.) above water level or the enclosure should be sufficiently slanted to provide both aquatic and terrestrial areas.

Northern frogs received between October and late March and to be used as active frogs may be stored in hibernation as described below (see Chapter V, Sections C.5 and 6). Northern frogs to be used for egg production and received during these same months *must* be stored in *hibernation.* Northern frogs received in other months cannot be used for egg production without provision for feeding and management designed to support oogenesis. Attempts should not be made to force northern "summer" frogs into hibernation unless the chilling is part of a specific, knowledgeable experimental design. They should be held as described above if they are to be used within a few days and as described below if they are to be held for longer periods.

Southern frogs must *never* be refrigerated or they will die. Attempts should not be made to obtain eggs from them until more is learned concerning their reproductive cycle. Southern frogs at *any season* of the year may be held temporarily as described above; if they are to be held for longer periods, they must be managed as described below for active frogs.

Active adults that are to be retained for more than a few days are best contained where the aquatic and terrestrial areas can be adjusted in accordance with the frog's known behavior. Such containers are described more fully in Chapter V, Sections C.4 and 6.

Water in the aquatic portion of the adult enclosures should be sufficiently deep to permit the animals to float in their typical position with eyes and nostrils above water, legs dangling and relaxed. Such a water depth should be provided even for primarily terrestrial animals; in the confinement of an enclosure, they obtain needed exercise by swimming rather than by jumping. This aquatic area also provides a suitable place for them to rest. Water should flow through the enclosure gently (two to three changes per day) but in sufficient volume to maintain bacterial counts below tolerance levels [see Chapter V, Section B.2.a.(7)]. Contrary to the circumstances for larvae, chlorinated water (4-6 ppm) for the reduction of bacterial counts may be used.

Although not as desirable as other types of enclosure (Chapter V, Section C.6), plastic vegetable crispers may be used. For these it is difficult to arrange flowing water and, if static water is used, caution is required because of the toxicity of some commonly used materials [see Chapter V, Section B.2.a.(10)]. A smaller opaque crisper with one end removed should be inverted inside a larger one to provide a terrestrial area; this constitutes a feeding platform on top and a hiding place below.

Matting made from a solid neoprene mesh should be placed on exposed surfaces. Not only does this keep the frog from direct contact with the plastic surface but also permits desirable air circulation about the animal.

Adult *R. pipiens,* especially from specific collection sites in the northern states, are highly susceptible to the Lucké renal adenocarcinoma. Whether as a result of stress or other factors, the development of this tumor is facilitated when the animals are held for long periods at room temperature under crowded conditions (Rafferty, 1962). Thus, experience suggests that 12 frogs are an appropriate number for a vegetable crisper measuring about 0.36 × 0.25 × 0.12 m (14 × 10 × 4.5 in.).

Aquaria or terraria can also be used. If the frogs are not in running water, the water should be changed frequently to prevent it from becoming fouled. Since frogs are normally secretive, several shards of unglazed clay flower pot placed in the container provide places to hide. Additional pieces of pottery in the water, but extending above it, provide a cool respite. Northern *R. pipiens* eat well and are active at 22–24 °C (72–75 °F); those from Mexico do best if held at higher temperatures, e.g., 25–27 °C (77–80 °F).

Optimal management principles to support and synchronize oogenesis in laboratory-reared and laboratory-bred *R. pipiens* are currently under investigation and as yet cannot be specified. However, a significant proportion of mature females will produce eggs if maintained as described above and fed as described below. No criteria have yet been established to recognize when eggs in living females are mature; however, mature eggs may be found 3–4 months following a previous ovulation. Such eggs will be resorbed within a few days of their maturity if the females are not artificially ovulated. This is also true of wild-caught gravid females if not returned to hibernation immediately. Pigment from such resorbed eggs is stored in the liver, which becomes intensely black as a result.

Preparations of active sperm have been made from male *R. pipiens* at all seasons of the year when they are managed following the procedures described here.

(e) Food for Adults   At metamorphosis *R. pipiens* shift from an omnivorous to a carnivorous diet comprised of food that must be moving. They differ from *R. catesbeiana* and *R. clamitans* in that they do not or cannot take food while submerged. Reports in the literature that "frogs and toads" will take food from mechanical devices such as "lazy susans" (Kaess and Kaess, 1960) do not apply to *R. pipiens.* A few individuals will strike at food presented in this manner, and an even smaller number will swallow it. Attempts are in progress to learn the criteria for motion, taste, and texture that must be met to present prepared foods to *R. pipiens.* At present, results do not warrant inclusion of a protocol for prepared foods in this document.

Adult frogs can survive for extended periods (3-4 weeks) without feeding if their quarters are clean, but long-term survival requires feeding the equivalent of 10-12 full-grown crickets two to three times a week. Frogs can be force-fed raw liver or hamburger, but not only is this time-consuming but quantity and timing are hard to judge and growth rates are poor. Therefore, when possible, use of live food is recommended. Definitive studies of *R. pipiens* dietary requirements have not been done because of the dependence on live food or force-feeding. Should devices for the presentation of prepared diets be successfully developed, such studies will be possible. At the University of Michigan Amphibian Facility, bone malformations that resulted from dietary insufficiencies have been observed. These have been alleviated, at least partially, by dusting the arthropod dietary items with commercial vitamin and protein supplements.

Some adult amphibians do not readily accept live food. These animals must be force-fed with special food mixtures from a syringe through a stomach catheter. Careful insertion of the catheter and a slow administration of the food will help prevent regurgitation. One ration that has been used successfully is prepared by homogenizing beef liver, eggs, and lettuce (all boiled) in a blender in a ratio of 4 : 1 : 1. A small amount of bone meal, one drop of cod liver oil/ml mixture, and commercial vitamin supplements may be added. Antibiotics may also be added in proper amounts (see Chapter IX, Section B), and the mixture frozen for later use. Canned dog food mixed with warm water and homogenized in a blender was another adequate force-feeding diet; when used for *Necturus,* cod liver oil was added (Kaplan and Glaczenski, 1965).

Food items from the wild, such as many arthropods and snails, can be used. One effective collection method is "light trapping" for beetles, moths, etc. Frogs fed on such diets, however, cannot be classified as laboratory reared or laboratory bred. To meet the criteria for these classifications, the animals must be fed living materials that have been reared in confinement (see Chapter III, Sections B.3 and 4), such as crickets, sowbugs, earthworms, beetles, flies, caterpillers, and moths. Even newborn or young mice will be readily taken. A word of caution must be added concerning the use of fly maggots: Because maggots of many fly species are not killed in the digestive tract of the frog, such maggots may destroy the frog. The problem of supplying a living diet, perhaps, has been the greatest single deterrent to developing laboratory-maintained cultures of *R. pipiens*; by proper selection of food types and standardization of food culture, however, this is not a prohibitively difficult problem and becomes increasingly feasible as the colony enlarges.

Many potential food items may be obtained commercially. Thus, although crickets and earthworms may be cultured, they may also be purchased from bait dealers and arrangements can be made for delivery on a regular basis. Sowbugs, which can be found in nature under old boards and dry leaves, are readily raised. They are recommended because they frequently appear on checklists of wild *R. pipiens* stomach contents, are "clean," and do not drown in the aquatic portion of the frog container. Indeed, they crawl out of the water onto the flower pots or up the sides of the container. Sowbugs may be raised at temperatures between 25 and 30 °C (77-86 °F) in roaster pans with a half inch of damp peat moss. The peat moss should be covered with a moist rectangle of corrugated cardboard and with a plastic lid from a vegetable crisper on the cardboard to reduce loss of moisture by evaporation. Sowbugs feed on crumbled rabbit chow and crushed blackboard chalk, which should be placed in a dry corner to retard molding. Both the rabbit chow and the peat moss bedding should be heat treated to kill organisms such as weevils. (The weevils not only compete with the sowbug colony but may actually destroy it.) Under these conditions, a colony of sowbugs will double its number about every 40 days.

Some frogs will kill themselves by overeating (e.g., the Japanese frog *R. nigromaculata*), but this does not occur with *R. pipiens*. Nevertheless, highly fed laboratory *R. pipiens* do develop muscle glycogen levels appreciably higher than those found in nature (Smith-Farrar, 1972).

(f) Hibernation   Hibernation is essential for holding Northern gravid females in the gravid state and is helpful in the winter months for low-cost holding of northern adults to be used for other purposes. Gravid females, if held under correct conditions, may be ovulated as late as July, 3 months after the time of normal ovulation. If correct conditions, however, are not met, the energy reserves in the eggs are mobilized and the eggs regress. Even a few days of improper conditions will lead to reduced fertility.

Summer frogs forcibly submerged without proper preparation for hibernation will drown. After entering hibernation, however, they remain submerged below the ice. During hibernation a respiratory shift from aerobic to anaerobic may occur to some extent, although cutaneous respiration does occur. With these factors in mind refer to Chapter V, Sections C.5 and 6 for descriptions of hibernation quarters. Note that for wild-caught northern *R. pipiens* water temperature should be between 3 and 4 °C (37-39 °F) in October and November, 1.5-2 °C (34-36 °F) from December, and 3-4 °C (37-39 °F) for 2-4 weeks prior to removal from hibernation, unless the frogs are being retained at room temperature only long enough to have ovulation induced.

Southern frogs should never be placed under hibernation conditions.

Criteria to recognize when an animal is in hibernation as distinct from cold torpor have not been devised. Such information is critically needed. Also, criteria for recognizing true emergence from hibernation are missing.

b. *R. catesbeiana*

(1) *Premetamorphosis*

(a) Enclosures   See Chapter V, Sections C.1-6 and Section B.1.a above.

(b) Environmental Controls   See Chapter V, Section B.

The management of embryonic and larval stages of *R. catesbeiana* is similar to that given for *R. pipiens* [see Section B.1.a(1)], except in rate of development and animal size.

Because of their large size *R. catesbeiana* tadpoles are placed in the frog enclosures as soon as the forelimbs emerge. During this stage mortality is reduced if the animal can move out of the water. The cause is not known but may be related to stress (see Chapter V, Sections C.2 and 3 and Section B.1.a of this chapter).

(c) Food Supply   The discussion of food supply for *R. pipiens* larvae is equally applicable here [see Section B.1.a(1)]. In addition to the prepared diet described, unflavored Knox gelatin has proven to be fairly effective as a binder at water temperatures below 24 °C (75.2 °F) (Culley and Meyers, 1972). The binding quality is best retained when the food retains its moisture. As more is learned about tadpole nutrition, undoubtedly new foods will become available.

The rabbit chow-gelatin mixture serves as an adequate food for several species of amphibian larvae. However, for *R. catesbeiana* this diet should be supplemented with a boiled leafy lettuce (not head lettuce). The reason is unclear, but growth appears to be better when lettuce is added. *R. catesbeiana* larvae from stock collected in the Gulf states have been successfully reared from egg through metamorphosis on the rabbit chow-gelatin-lettuce diet in 3-4 months at 30 °C (86 °F). At metamorphosis, the larvae weigh about 10 g (0.35 oz).

(2) *Postmetamorphosis*

(a) Rearing Facilities   The most successful rearing containers for the frog stage are those that have a wet and dry area and a continuous supply of fresh water. Crowding of adults creates no serious problems as long as sanitation is maintained. Juvenile bullfrogs placed in containers with as little as 120-150 mm² (1.86-2.33 in.²) of floor area per frog have been successfully reared for over 2 years with the same degree of crowding. Specifications for enclosures are given in Chapter V, Sections C.3, 4, and 6.

(b) Environmental Controls   The same water quality criteria holds for frogs as for tadpoles (Chapter V, Section B.), with the exception of chlorine. As long as chlorine is less than 3-4 mg/liter, the bullfrogs are not affected. Long-term exposure to concentrations above 4 mg/liter may be detrimental. Since chlorine in water supplies is occasionally as high as 4 mg/liter at the tap, the supply should be checked periodically. Although wide spectrum light is recommended for laboratory-reared frogs, bullfrogs maintained for over 3 years under fluorescent lights have shown no signs of vitamin D deficiency (Culley, 1973).

(c) Food Supply   The diet of wild southern bullfrogs consists mainly of crayfish, supplemented with fish and, at times, large quantities of tadpoles. Given a choice, however, bullfrogs prefer tadpoles and fish. Crayfish are nutritionally adequate for bullfrogs, but have much less food value than fish of the same weight.

Tadpoles of either *R. catesbeiana* or *R. pipiens* may be used as food for bullfrogs; however, these tadpoles must be of a size that the frog can swallow, a factor of particular concern when feeding juvenile bullfrogs. Though wild-caught tadpoles may be used, they may affect the classification category of the frog (see Chapter III, Section B.3).

Although tadpoles appear to provide an adequate diet for short-term feeding, a long-term diet of bullfrog tadpoles for young bullfrogs may not be nutritionally adequate. Tests have shown that after 2 months young bullfrogs on this diet feed erratically, become sluggish, and lose their skin brilliance.

The use of crickets, mealworms, or earthworms as single diets or in combination have not been nutritionally adequate. In recent studies at the Louisiana State University (LSU) facility, bullfrogs on this diet developed bone deformations within 3 months.

Thus, only fish have proven to be nutritious and accepted on a long-term basis. Several bait minnows have been utilized in feeding bullfrogs, although it must be cautioned that using minnows raised in outdoor ponds may affect the standard classification of the frogs (see Chapter III, Sections B.2 and 3). Using growth as the criterion, minnows, so far, have proven to be nutritionally adequate. Two species of small minnows—mosquito fish (*Gambusia affinis*) and sailfin molly (*Mollienisia latipinna*)—have been used at LSU. However, both become infested with a nematode (*Eustrongylides wenrichi*) that attains a length of 70-100 mm (3-4 in.). When infested fish are consumed, the nematodes burrow through the stomach wall of the frog and migrate randomly. Many locate just under the skin and in the muscle tissue, apparently causing little harm to the frogs. Others move into the liver, kidneys, or heart and death results. The nematode has been found only during the summer and early fall in the fish in the LSU area. It is advisable, therefore, not to use these fishes as a source of frog food, particularly during the seasons indicated. So far this nematode has not been found in goldfish (*Carassius auratus*), fathead minnows (*Pimephales promelas*), or the golden shiner (*Notemigonus crysoleucas*)—common bait minnows that have been used for feeding frogs.

One of the most successful regimes for bullfrogs has been a combination of crickets, earthworms, and golden shiners fed daily. Although inadequate when used alone, worms and crickets have been successfully used to supplement the nutritionally adequate shiners and results in some cost reduction. Where shiners have not been available, excess live juvenile and adult mice and hatchling chicks and quail from the animal facility have proved to be an adequate substitute.

Because bullfrogs feeding under water have difficulty catching their food if the water is deeper than 10-20 mm (0.39-0.78 in.), a depth at which fish can escape, water depth in their enclosures must be controlled. If water depths are not optimal and if different-sized frogs are in the same enclosure, cannibalism may result. As long as food is abundant and readily obtained, cannibalism seldom occurs.

For frogs under 2 months of age, the water depth should not exceed 10-20 mm (0.39-0.78 in.) and not cover over one fourth of the floor. Excess food should be added to facilitate capture, and dead food should be removed daily. These young frogs will not be very successful in taking live food over 20 mm (0.78 in.) in length. After 2 months growth, food up to 40 mm (1.56 in.) can be captured, and after 4 months food size is not critical, except for the slower growing frogs.

(d) Hibernation   Little information is available on hibernation requirements for *R. catesbeiana*. However, techniques now employed with *R. pipiens* may be applicable to *R. catesbeiana* collected in the North. [See Chapter V, Sections C.5 and 6 and Section B.1.a(2) of this chapter.]

To keep southern *R. catesbeiana* under hibernating temperatures for periods up to 6 weeks, 10 juveniles or 5 adults can be placed in nylon netting sandwiched between moist layers of sphagnum moss in a 3-gal plastic container. The frogs are placed in these containers at ambient air temperatures and then stored under refrigeration. The container will cool to temperatures of 5-7 °C (41.0-44.6 °F) in about 24 h. In this temperature range mortality will not occur if 5-7 °C (41.0-44.6 °F) water is added to the container for a few minutes every 2 weeks and then drained off.

Wood excelsior or peat moss is inadequate as a substitute for sphagnum moss. The wood excelsior develops heavy fungal growth and the peat moss packs too tightly.

The animals should be removed slowly from storage at low temperatures by placing the container at room temperature for approximately 24 h. The only obvious detrimental effect noted has been the development of temporary skin lesions on the webbing within 48 h after removal from refrigeration.

c. Other Ranid Species

The care of other ranid species encountered in American laboratories follows either the procedures for *R. pipiens* or *R. catesbeiana*, depending on their affinities or behavior patterns. *R. grylio* and *R. clamitans* resemble *R. catesbeiana* in their habits and may be handled following the procedures for *R. catesbeiana*. *R. palustris*, which may be hybridized in the laboratory with *R. pipiens*, may be treated as *R. pipiens*.

R. sylvatica is, as the name implies, a woods frog. It is very secretive and spends little time in the water except to breed. All indications are that they hibernate, not in water, but burrowed under the forest litter. Consequently, they should be kept in containers with a minimum of water, a maximum of terrestrial space, and appropriate places to hide (unglazed flower pot shards). *R. sylvatica* are difficult to maintain in the laboratory, especially during the period from metamorphosis to sexual maturity. Further studies are needed to identify the reasons for these difficulties since a similar species, *R. japonica*, has proved to be highly adaptable to laboratory conditions in the Laboratory for Amphibian Biology of the University of Hiroshima, Japan (Kawamura and Nishioka, 1972).

In addition to the divisions that have been noted above in describing these several species, they can be further divided into two categories: those that breed immediately upon emerging from hibernation and those that mate following a period of nutritional intake between hibernation and breeding. The former include *R. pipiens, R. palustris, R. sylvatica*; the latter, *R. catesbeiana* and *R. clamitans*.

Artificial breeding of northern representatives of the first group is readily accomplished through much of the year because they are gravid at the time of entering hibernation and may be bred at any time when removed from hibernation (see Chapter VII). For the latter group, however, this is not yet possible. Additional study of endocrinology and nutrition is needed to bring the artificial fertilization of these species under control (Sarkar and Appaswamy Rao, 1971).

The conditions for hibernation of the several species show some differences, especially for those species with extended latitudinal distribution. Representatives collected north of the line separating ice-free from ice-covered ponds may be maintained as described for *R. pipiens*. Animals collected south of this line have not been critically tested. *R. nigromaculata*, which has been studied in Japan (Kawamura and Nishioka, 1972), is native to a territory where ponds seldom freeze. They burrow in the dikes between rice paddies and under forest litter. Hibernation or dormancy temperatures for these animals are between 7 and 10 °C (45-50 °F). Presumably, American animals collected from similar climatic areas could best be maintained under similar temperature conditions (see Chapter V, Section C.5).

2. Other Anurans

a. *Xenopus*

The care of *Xenopus* has been extensively described (Weiz, 1945a,b; Nieukopp and Faber, 1956; Frazer, 1966; Gurdon, 1967; Brown, 1970; Deuchar, 1972). However, a brief review may be useful.

(1) *Larvae* Development of *Xenopus* is rapid. At 20-22 °C (68-72 °F) cleavage and gastrulation take place within 1 day. Hatching occurs on the second and third day.

The larvae should be fed when they begin to swim along the bottom at an angle of about 45° with their heads down and tails up. Gasche (1943, 1944) introduced powdered nettle as the food for larval *Xenopus*. Although good, it is not necessary. Dried green pea soup or other finely ground food is satisfactory. The food is mixed with water and *allowed to settle*, and the supernatant containing fine food particles is decanted. The decanted suspension is delivered to the tadpoles in measured amounts. Care must be exercised as coarse food particles congest the filter-feeding mechanism of these larvae and can produce death (Weiz, 1945a,b). The quantity of food administered should permit clearing of the water by the tadpoles about 4-5 h after feeding. A fresh food slurry should be made each day; since growth is facilitated by a second daily feeding, the same slurry can be utilized a second time on the same day.

The larvae may be transferred from the amplexus chamber (see Chapter VII, Section A.10) to larvae enclosures after they have fed for a few days. They should be siphoned or dipped but not netted until they have grown for about a week. Plastic vegetable crispers are satisfactory as larval enclosures if the density of the larvae does not exceed 6–8 per liter. The water should be changed and the enclosures cleaned at 3-day intervals. *Xenopus* larvae can also be raised in the flowing-water bottle system used for ranid larvae if the screen in the neck of the bottle (see Chapter V, Section C.6.a) is covered to retain the suspended food.

Under the conditions of temperature, feeding, and water changing described above, metamorphosis should begin 5–6 weeks following fertilization. Metamorphosis, itself, requires 15–20 days. Shortly after forelimb eruption, the larvae stop feeding. This is frequently first detected by the failure of the animals to clear the water after feeding. After their tail is half resorbed, the animals will eat finely diced meat, mosquito larvae, *Tubifex* (redworms), etc. A suitable diet has been suggested made of ground fat-free beef heart and powdered milk blended together to yield a dry, crumbly material. This may be used as steady diet, whereas *Tubifex* or mosquito larvae used alone will result in poor growth or death of the young adults, presumably as a result of nutritional deficiency. Care must be taken to remove uneaten food and to keep the water clear. After a month or so the juvenile animals may be treated as adults.

(2) *Adults* Adult *Xenopus* are maintained in aquaria or other suitable enclosures with a volume adequate to the numbers of animals (one animal/2 liters) (see Chapter V, Section C.6). The walls of the enclosures should be sufficiently high or the enclosures should be covered to prevent escape. In their natural habitat, the frogs are subject to considerable temperature variation. Thus, although they can tolerate tepid water, water at normal laboratory temperature is suitable. The water must be dechlorinated and should be at least 0.15 m (6 in.) deep. It is convenient to equip the enclosures with an overflow pipe to permit constant gentle flow of the water or the water should be changed periodically to prevent stagnation [see Chapter V, Section D.2.a(10)]. Fecal and food wastes must be removed within several hours after each feeding, either by draining the enclosure or by using an aspiration device.

Though a liver diet has proved satisfactory for maintaining *Xenopus*, experience indicates that pieces of beef heart cut to resemble earthworms of a size appropriate to the size of the animals being fed is a superior diet. Chunked or diced meat is unsatisfactory because the poor surface–volume relationship does not allow adequate digestion. The pieces of beef heart should be soaked in a commercial vitamin mixture. Earthworms are a highly effective food. Feeding may be as infrequent as twice a week although more frequent feeding results in more rapid growth.

b. *Bufo*

Bufo are more terrestrial than ranids (Blair, 1972). Enclosures with shelves spaced appropriately to the size of the animals can accommodate large numbers. Such enclosures are described more fully in Chapter V (Sections C.4 and 6.a) and in Frazer (1966). Water need be provided in only modest quantities; as a trickle on the bottom of the cage or in a Petri dish or other shallow container. If the floor of the enclosure is moist, shelving or low tables covered with matting, such as neoprene, should be used to provide the animals a dry area.

The food for *Bufo* may be highly varied. Crickets, sowbugs, earthworms, or any of the items mentioned in the diet for *R. pipiens* may be used. Mealworms (*Tenebrio*) are appropriate for toads because they will not drown in the minimal amount of water in the containers. In addition, some toads will adapt to nonliving diets: *B. marinus* has been maintained on canned cat food or wet dog food. Force-feeding of liver and fish proteins has not proved satisfactory (Jakowska, 1972).

c. *Bombina orientalis*

These animals are climbers rather than jumpers and will escape if their containers are not carefully closed. Plastic or enamel pans, 50–76 mm (2–3 in.) deep, securely closed with nylon or metal screening are suitable for housing and are necessary if flying insects are used for food. The floor of the pan should be lined with neoprene matting on which shards of unglazed flower pots are placed. The water need only cover the matting but should be flowing or be changed every second or third day to prevent the accumulation of toxins secreted by the animals. This toxin constitutes only minimal danger to laboratory personnel; the irritation produced on mucus surfaces simply guarantees that the worker will wash after handling them (see Chapter X, Section B.2).

The animals may be maintained throughout the year at laboratory temperatures. They show little preference for different types of food provided the food is not too large: i.e., crickets, earthworms, sludge worms, sowbugs, and flies [not maggots, see Section B.1.a(2) this chapter]. Crickets and sowbugs are recommended because they are available over a full series of graded sizes and can thus be selected appropriate to the size of the animal. Fed and manipulated as described in Chapter VII, these animals can be induced to ovulate every second day, although in nature, they seem to mate only once a year.

C. URODELA

1. Axolotls (Neotenic Larvae of *Ambystoma mexicanum*)

The following description is modified from a privately distributed instruction sheet that was prepared by R. R. Humphrey of Indiana University, Department of Zoology, Bloomington.

a. Early Larvae

Axolotl larvae do not require food immediately after hatching since at that time the stored yolk has not all been metabolized. Feeding should be initiated when the color of the intestine changes from white to a gray or gray-black, or shows a slightly greenish tinge from bile accumulation. Larvae that may have begun to float before feeding, because of air in the stomach, should at once be put in very shallow water and generously fed.

Recently hatched brine shrimp (*Artemia*) are among the best food with which to begin feeding axolotls. Mosquito larvae are also excellent. Experience at The University of Michigan Amphibian Facility demonstrates that mosquito larvae are readily produced, and because they may be harvested at any of their developmental stages, they are available in all sizes suitable for use as food throughout the lifetime of the axolotl larvae. *Daphnia* of small size or very small *Tubifex* or *Enchytraeus* may be used. These worms tend to "ball up" in masses but can be reduced to short lengths by cutting into the mass several times with scissors. Axolotl larvae at hatching are smaller than larvae of some other amphibian species; if the food first given is too large for them to handle, they may fail to start eating and die of starvation. Once they begin eating, they should receive only as much food as will be consumed within a few hours. Before feeding again, change the water to remove feces and uneaten food. Dishes require frequent scrubbing to keep them free of protozoa.

Neither brine shrimp ncr *Enchytraeus* alone constitute a suitable food for indefinite use. Larvae fed nothing else eventually show edema and/or ascites and hemorrhages in the limbs, skin, or wall of the stomach and soon die. Those that survive may show serious adhesions of the viscera. To prevent loss of animals from dietary insufficiency, add *Daphnia* or *Tubifex* to the diet, or earthworms cut into short pieces. Smaller amphibian larvae, embryos, or even infertile eggs (removed from membranes) may also be fed. Hand feeding with beef liver cut into very thin, narrow strips, while time-consuming, will save larvae from dietary insufficiencies if begun sufficiently early. Even when a variety of live food is available, the larvae should ultimately be hand fed with liver if they are to be brought to breeding condition when one year of age, unless maximal feeding by other methods is possible and has been found to be equally effective. Daily feeding for young larvae is advisable. After several months or as the animals approach maturity, this may be discontinued and food provided only on alternate days. It may be helpful for the welfare of the animal (growth and mortality levels) to add a small quantity of bone meal to the diet occasionally; sprinkle the meal over the sliced liver and stir to cause the meal to adhere to it.

Young axolotls are seldom troubled by mold. If they become infected, however, they may be placed for several hours in a weak solution of Mercurochrome (1 : 500,000 or weaker). This treatment can be repeated if necessary.

Larvae infected with a disklike protozoan (probably of the genus *Trichodina*) have been successfully treated by immersing them for a few seconds in a strong solution of Mercurochrome (a deep red color) and rinsing them under a low-pressure tap flow. Richardson (1937) reports successful treatment of similarly infested young trout by a 2-min immersion in 3 percent saline or a 15-sec immersion in 1 : 1,500 glacial acetic acid.

b. Mature Larvae

These larvae, which reproduce without undergoing metamorphosis, are totally aquatic and must be treated accordingly.

(1) *Enclosures* Axolotls do not require running water or special provision for the aeration of water, provided the enclosures are of suitable size and are kept clean. The water should be changed after feeding and whenever it becomes fouled, and the containers should occasionally be washed with a detergent to keep them free of bacteria and protozoa. The siphoning device described for use with *Xenopus* can also be used to remove debris from axolotl containers. Although several animals may be put together in a large aquarium, it is better to keep them separated, either in smaller aquaria or in glass or plastic bowls. This facilitates the breeding regimen and avoids the cannibalism that may occur if they are inadequately fed. Glass fish bowls of one gallon capacity are adequate, although larger ones may be provided for old animals of maximum size. Those of quart or 2-quart size are useful for smaller animals. Rectangular plastic containers instead of bowls are used in some laboratories and have the advantage of better stacking. Other enclosures are presented in Chapter V, Sections C.6.a and d.

Axolotls must be provided with chlorine-free water (see Chapter V,

Section B.2.a). Highly aerated water should be avoided since it will result in reduction of gill arborization and, while not a handicap to animals maintained under these conditions, it may result in difficulties if the animals are later placed in poorly oxygenated environments as during shipping. Their water should be maintained at 20-22 °C (68-72 °F). Colder water will cause the animals to reject food or to regurgitate it if they have been recently fed under warmer conditions. Warmer water is detrimental to breeding. Particular care should be exercised in adjusting the temperature of fresh water when the enclosures are cleaned.

(2) *Feeding* Adult axolotls can be fed earthworms or many types of insects. However, beef liver, lean beef, or beef heart may be used. Cut into pieces suitable for a single feeding, the meat can be frozen. In preparation for feeding, the meat is cut into thin slices while frozen or partially thawed. These, in turn, are subdivided into narrower strips; earthworm size for large animals and toothpick size for smaller animals. Pieces cut to provide maximum surface area–volume ratios will facilitate digestion. The food should be handled with blunt forceps and offered to each animal as long as it is readily accepted. Animals heavily overfed or *given large thick pieces* are likely to regurgitate. Axolotls do not readily digest either fat or collagenous tissue (such as fascia or tendon); these should be trimmed away, as should parts of the very large blood vessels in liver. The animals seem to prefer liver to other food; beef and lamb liver are both suitable, but pig liver is frequently regurgitated. Since their food is digested rather slowly, adults need be fed only on alternate days or on a schedule of three feedings/week.

2. Other Urodeles

Little is known about the laboratory management of the larval stages of other salamanders. Generally, they should be maintained in well-oxygenated, clean, chlorine-free water. Most are carnivorous and will eat live food such as mosquito larvae, Encytrea worms, brine shrimp, and other small crustaceans and annelids. Most species should be maintained at 10-20 °C (50-68 °F).

Metamorphosed salamanders may be maintained in terraria or other containers that provide moist hiding areas and access to free water. The containers described above for adult *R. pipiens* are suitable. Live insects and earthworms are usually accepted as food.

Necturus should be maintained in clean, filtered, and well-aerated water. Their containers should provide a gravel bottom and shelter, such as broken pot shards or flat stones, where the animals may hide with their backs against the overhead shelter. Normally, they will feed on earthworms and small crayfish and will sometimes accept small strips of raw meat or small dead fish after they have fed on live food while in captivity.

Notophthalmus viridescens efts and adults may be maintained in a semi-aquatic terrarium or container similar to that described for *R. pipiens* and fed on earthworms, mealworms, crickets, and other small insects. The adult newt stage can also be maintained in aquaria or similar tanks, but a semi-terrestrial terrarium is recommended.

VII Breeding

A. ANURANS

1. General Comments

Species that normally mate immediately upon emergence from hibernation can be readily artificially inseminated any time during hibernation. Laboratory-reared or laboratory-bred specimens may be in a physiological state allowing ovulation at any time. Artificial insemination using species that normally mate some weeks after hibernation has been accomplished only in the days immediately before the time of natural breeding. Laboratory-maintained specimens of other species, such as *Xenopus, B. orientalis* (The University of Michigan Amphibian Facility information sheet), and *Engystomops pustulosus* (Davidson and Hough, 1969), may be ovulated at any time depending on the physiological cycling regimens followed.

a. Sexing

See Chapter II.

b. Readiness for Reproduction

See Chapter VI, especially Sections B.1.a(2) and c.

c. Identification of Individuals

See Chapter VIII.

2. Artificial Induction of Ovulation

The following description is based on *R. pipiens*, whose ovulation has been most routinely conducted under laboratory conditions (Rugh, 1965;

TABLE 8 Pituitary and Hormone Doses for Inducing Ovulation

Month	Pituitaries Alone	OR	Pituitaries	+ Progesterone (mg)
Sept–Oct	10–12		2	5.0
Nov–Dec	6–8		2	2.5
Jan–Feb	4–5		1	2.5
Mar–Apr	1		1	–

Di Berardino, 1967). Kits including pituitary preparations plus directions for fertilizing eggs are available from commercial dealers. However, pituitaries are readily collected following the technique described by Rugh (1965). These may be used immediately or following storage at 4 °C (39.2 °F) in absolute (not 95 percent) ethanol or following lyophilization. Since seasonal changes in pituitary potency and in female sensitivity occur, Table 8, based on the use of pituitaries collected at the seasons indicated, will serve as a guide to dosages (see Bagnara and Stackhouse, 1973). The technique was developed by Wright and Flathers (1961). Masui (1967) and Schuetz (1967) discuss the pertinent physiological mechanisms.

Doses shown in Table 8 are calculated as female pituitaries; two male pituitaries are equivalent to one female pituitary. Both the pituitary and progesterone* should be injected into the coelomic cavity about 48 h before scheduled fertilization.

3. Amplexus

If breeding is to be accomplished by amplexus, both male and female should be treated. Follow the above table for pituitary dosage for females. Males will respond to half the pituitary dose; progesterone is not needed. The injected animals should be placed in an aquarium with 200–250 mm (8–10 in.) of dechlorinated water at 20–22 °C (68–72 °F) (see Chapter V, Section B.2.a). The animals should be placed above plastic screening with a coarse mesh 9.5 mm (3/8 in.) held approximately 50 mm (2 in.) above the floor of the aquarium to protect the egg masses. Depending on the season, fertile eggs should be available in 18–48 h if the aquarium is placed in an undisturbed area. Note that the precise time of fertilization is not known when this technique is used.

---

*Progesterone may be obtained from organizations such as Nutritional Biochemicals Corporation and Sigma Chemical Company.

4. Artificial Insemination

Eggs may be obtained for insemination following the technique described above for artificial induction of ovulation. Sperm should be prepared when eggs are observed at the cloaca following the application of slight pressure to the abdomen of the female. The male need not be injected with hormone for this purpose.

To obtain the sperm, pith a male frog, remove both testes, dissect away adhering tissues, and wash the testes free of blood. Macerate them in 10-20 ml of medium consisting of pond water, dechlorinated water, or 10 percent Steinberg's solution [see Chapter VI, Section B.1.a(1)]. A convenient way to do this is to force the testes through a syringe with an 18-gauge needle. Large clumps may be broken up by aspiration. Wait 15-20 min for full sperm activity, which can be determined by examination under a compound microscope. This preparation containing sperm from two testes may be brought to 100 ml prior to inseminating the eggs. Higher concentrations may be used but young (October for wild-caught frogs) or old (artificially prolonged hibernation) eggs are susceptible to polyspermy when dense sperm preparations are used.

Sperm may also be obtained from *R. pipiens* without the death of the donor by injecting males with 100 IU of human chorionic gonadotropin (HCG). Following this injection, the animal should be held in a quiet area and any manipulation should be gentle to avoid premature urination. Sperm of maximum activity may be recovered by stimulating urination after 1-3 h (McKinnell, 1962). The urine that contains the sperm should be used as described below.

Where sperm of rare or genetically defined males are to be used, destruction of the male is to be avoided. Where hormone stimulation of sperm release has been ineffective, single testes or portions of testes may be surgically removed. Using the appropriate anesthesia (see Chapter IX, Section F), a testis may be exposed through a 5-mm (0.2 in.) dorsolateral incision and part of or the whole testis may be removed. No procedures for hemostasis are needed and the incison may be closed with simple surgical stitches.

After sperm activity has been confirmed, eggs are stripped from the female (Figure 23) by holding her in such a manner that pressure is applied to her abdomen with the force directed toward her cloaca, thus squeezing the eggs from the ovisacs into a dry Petri dish. This is accomplished by bringing the legs of the female forward parallel to the abdomen. This assures that pressure is not dissipated laterally. The third and fourth fingers are applied firmly over the throat and thoracic region to avoid dissipation of pressure anteriorly. The middle finger, index finger, and thumb

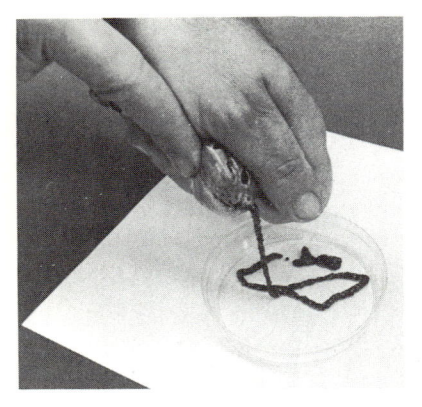

FIGURE 23 Expressing eggs from a female *Rana pipiens*.

are used to "milk" the abdomen toward the cloaca. Initially, rather sharp pressure may be needed to open the cloacal sphincter muscle; subsequently, gentle pressure will suffice to aid the natural peristalsis. This method requires the use of only one hand; the other is free to wipe moisture from the cloaca so that eggs do not stick to the skin, to manipulate the collecting dish, or to assist in the "milking" action. If the tongue of the female protrudes during this process, pressure is being applied incorrectly.

"Milking" eggs from large frogs, such as bullfrogs, may be aided by wrapping the animal with an "Ace bandage." Start the wrap snugly at the throat and wind spirally, loosening the wrap as it progresses over the abdomen.

The eggs expressed into the Petri dish are inseminated by pipetting the sperm suspension over them, seeing that each egg comes in contact with the sperm preparation. Just moistening the egg strings with the sperm suspension conserves sperm. This application of sperm must be done before the egg jelly has started to swell. After 10–15 min flood the eggs with medium. Use of a dry Petri dish is recommended because the eggs stick to its surface. This facilitates changing the medium. A spiral pattern is formed in the Petri dish to prevent the eggs from clumping. After an additional 15 min pour off the medium plus sperm and replace with fresh medium. Set the eggs aside for 30–40 min. They may then be *scraped* free from the Petri dish and transferred to culture bowls or trays. A single-edge razor blade, tissue section lifter, scalpel, or glass slide should be used for this: Do not attempt to pull the eggs from the Petri dishes with fingers or forceps as they may be distorted and thus damaged. About 1 h after fertilization those eggs successfully fertilized will rotate so the black animal hemisphere is uppermost.

Using scissors, cut the clutch of eggs into masses of 10–25 eggs each. This allows escape of waste products and sufficient surface for gas exchange. In nature the volume and movement of water in the ponds adequately handle these functions. The eggs are now ready to be treated as described for the stages of early development (Chapter VI, Section B.1.a).

The procedure for artificial insemination described here is applicable to any type of mating involving biparental reproduction (e.g., random mating lines, heterozygous marked lines, mutant lines) (see Chapter III, Section C).

5. Parthenogenetic Lines and Haploid Animals

For parthenogenesis, eggs are obtained as described above and, depending on the specific objectives, are inseminated and activated with ultraviolet-irradiated (Nace *et al.*, 1970) or toluidine-blue-treated (Briggs, 1952) sperm of the same species but carrying a dominant mutation as a marker (e.g., Burnsi sperm on wild-type eggs) or of species whose normal sperm would produce lethal hybrids (e.g., *R. clamitans* sperm on *R. pipiens* eggs) (Moore, 1955); such eggs can also be activated by pricking with a needle (Shaver, 1953). After 20 min at 18 °C (64 °F), the inseminated eggs are exposed to 37 °C (99 °F) for 4 min and returned to 18 °C (65 °F). In a significant percentage of eggs this heat shock results in retention of the second polar body and produces diploid animals whose genome is of maternal origin. This particular form of parthenogenesis is called gynogenesis because it utilizes only the maternal genome. It is used for the production of the inbred and gynogenetic diploid lines (see Chapter III, Sections C.1.c and f). It is modified for haploid animals (Chapter III, Section C.1.h) by eliminating the heat shock.

The irradiation used to treat the sperm is from a 15-W Westinghouse Sterilelamp (G15T18) at a distance of 0.84 m (2.76 ft) for 5 min. Sperm thus treated retain their motility and are capable of insemination; only their nuclei are inactivated. Toluidine-blue-treated sperm are similarly inactivated. Should sperm inactivation be ineffective, the dominant phenotype will appear among the progeny when the homologous mutant sperm are used, or death of the progeny will occur as a result of hybrid incompatibility when foreign sperm are used. Should the heat shock have been ineffective in supressing meiosis II, the progeny will develop the haploid syndrome and die.

The technique just described is suitable for those species that lay their eggs in cold water shortly after emerging from hibernation. For those species that lay their eggs after a period of activity and when water in the breeding sites has warmed, cold shock should be substituted for the heat shock. The conditions for ultraviolet irradiation and the precise timing and the temperature of either the heat or cold shock is dependent on the species and perhaps also on geographic variance. Thus the investigator should confirm or evaluate these values in each laboratory and for each species.

The earliest studies of parthenogenesis were of eggs stimulated to develop by agents other than sperm (Loeb, 1899). The cleavage-initiating factor responsible for the success of parthenogenesis in *R. pipiens* is a protein (Frazer, 1971).

6. Homozygous Lines and Androgenesis

Because of crossing-over, three generations of gynogenesis are required to attain 99 percent homozygosity (Nace *et al.*, 1970). Absolute homozygosity may be attained in one generation, however, by using the insemination techniques described for gynogenesis and eliminating the diploidization step. Thus development is started as though haploid animals were to be produced. However, when the first cleavage furrow becomes evident, the eggs are exposed to 5,000 psi for approximately 4 min using a hydrolic press. This step inhibits cytokinesis but allows karyokinesis to continue, thus diploidizing the eggs. Since the entire diploid genome derives from the post-meiosis II haploid set, total homozygosity is attained. Animals obtained by this procedure are poorly viable unless derived from a selected or inbred or gynogenetic line.

Androgenesis is development utilizing only the paternal genome. The maternal nucleus of an egg is destroyed by irradiation (Gurdon, 1960; McKinnell *et al.*, 1969) or by mechanical removal after insemination by sperm of the same species (Porter, 1939) but before karyogamy. Upon the initiation of the first cleavage, pressure is applied resulting in homozygous androgenetic development. Such animals are as labile as maternally derived homozygous animals.

Homozygous diploids can be produced by the nuclear transplantation method (see Chapter III, Section C.1.g and Chapter VII, Section A.9). These homozygous diploids, as other homozygous individuals, show reduced viability (Subtelny, 1958).

7. Polyploid Animals

Polyploid animals either occur spontaneously (Fankhauser, 1955; Kawamura and Nishioka, 1963, 1967) or are produced by following normal biparental insemination or gynogenetic insemination by first cleavage inhibition (see Chapter III, Section C.1.i). Triploid animals are most frequently produced by inhibiting the second meiotic division following normal biparental insemination.

Polyploid animals occur in two types of nuclear transplantation experiments. In the first type, a somatic nucleus is fused with the maternal gamete nucleus or with the zygote nucleus (Sambuichi, 1959; Subtelny and Bradt, 1963; McKinnell, 1964). In the second type of experiment, spontaneous delay of one cytoplasmic cleavage interval, accompanied by karyokinesis at its proper time, results in the production of embryos of double the expected chromosome number (Gurdon, 1959).

8. Mosaic Animals

Application of the pressure technique to inhibit cytokinesis at the second or third or combination of cleavage stages can produce mosaic animals. Depending on which of the previously described techniques were used to initiate development, a variety of mosaic types can be produced.

9. Nuclear Transplantation

Nuclear transplantation was first successfully accomplished in vertebrates by Briggs and King (1952). The procedure has been described for the leopard frog by King (1966, 1967), *X. laevis* by Elsdale *et al.* (1960), and urodeles by Signoret *et al.* (1962).

Activated eggs are enucleated by manual removal (Porter, 1939), ultraviolet irradiation (Gurdon, 1960), or laser irradiation (McKinnell *et al.*, 1969). Donor cells are dissociated in an electrolyte medium that may contain a chelating agent and/or a protoeolytic enzyme (King and Briggs, 1955). A dissociated cell is drawn into a micropipette in such a manner that its plasma membrane is broken. This liberates the undamaged nucleus. The micropipette containing the nucleus and some protective cytoplasm is then inserted into the previously activated and enucleated host egg. Low temperature and a polycationic amine enhances certain nuclear transfer experiments (Hennen, 1970). As stated elsewhere, these procedures may be used to produce isogenic groups (see Chapter III, Section C.1.b), polyploids (see Chapter III, Section C.1.i), and homozygous diploids (see Chapter III, Section C.1.g).

10. *Xenopus*

Gurdon (1967) has described breeding and husbandry methods for *X. laevis*. Eggs may be obtained from female *Xenopus*, judged to be gravid by abdomens distended with eggs and by their somewhat reddened cloacal lips. Though not absolutely necessary, sperm are best obtained from males with dark "nuptial pads" on the underside of their forearms. In laboratory-maintained colonies ovulation may be induced once a week for several weeks (Wolf and Hedrick, 1971). Both female and male are stimulated to breed with chorionic gonadotropin, which is usually dissolved in physiological saline at 500 or more IU/ml. Injection is into the dorsal lymph sac. The dose administered depends on the size of the animal; unless the animals are quite small, however, a total of 500 IU per female and 250 per male is usually satisfactory. Best results are obtained when infrequently ovulated females are administered 100-unit "primer" doses about 5 h before administering the remainder of the quantity. Two males should be prepared for each female and be kept separate between the "primer" and final injection. After amplexus has been established, the unsuccessful male should be removed.

The fully injected animals are placed in an amplexus chamber as described in Section A.3 above for *R. pipiens*. The temperature of the medium should be 20–25 °C (68–77 °F). Below 20 °C (68 °F) the likeliness of success is lessened. The amplexus chamber should be covered and placed in a quiet location to minimize distraction. Spawning should begin within 12 h and continue through the next 24 h. The parents are then removed and the larvae separated from the screen of the amplexus chamber; hatching occurs between 2 and 3 days after the spawning. The larvae should be left in the amplexus chamber until after they have fed for several days. They may then be transferred to containers as described in Chapter VI, Section B.2.a(1).

Xenopus eggs may also be artificially inseminated. Wolf and Hedrick (1971) developed the technique quoted below:

Procurement of Gametes

Oviposition began 6–12 hours after the administration of hormone, and eggs were obtained by stripping laying females in a manner similar to that described for *Rana pipiens* (Rugh, 1965). Due to the limited capacity of the ovisac of *Xenopus laevis*, only 400–1000 eggs could be stripped from the animal at a time; however, additional eggs could be obtained upon repeated stripping every 1 to 2 hours.

Sperm suspensions were prepared by macerating excised testes in 0.05 DeBoers solution with a glass rod in 12-ml conical centrifuge tube. DeBoers solution (DB) consists of 0.11 M NaCl, 0.0013 M KCl, 0.00044 M $CaCl_2$ with the addition of $NaHCO_3$ to pH 7.2 (Katagiri, 1961). The concentration of DB employed is expressed in decimal form. Other standard salt solutions such as Amphibian Ringers (for composition see Rugh, 1965) can be used in place of DB. This solution is approximately the same ionic strength as DB, one of the most important parameters affecting sperm motility and viability (see results). The administration of human chorionic gonadotropin to mature males resulted in the appearance of black nuptial pads on the inner aspects of the forelimbs but did not noticeably affect the viability or fertilizability of the resulting sperm suspensions and, therefore, was not routinely employed. Sperm counts and motility were determined at 100-fold magnification with a light microscope. Only active swimming spermatozoa were classified as motile. Estimates of sperm concentration were made from hemacytometer counts and the average value of 25×10^6 cells/ml was obtained for sperm suspensions resulting from the maceration of one pair of testes in 10 ml of buffer.

Artificial Insemination

Artificial insemination was conducted in watchglasses (3–4 inch diameter) by stripping eggs (50–100) directly into sperm suspensions prepared in 0.05 DB or by adding a concentrated sperm suspension (0.05–0.1 ml) to eggs (10–500) in 0.05 DB. The insemination mixture was agitated manually for several minutes to ensure uniform exposure, and 10 minutes later samples were flooded with 0.05 DB. [The sperm suspension should be used immediately after preparation at room temperature or it should be iced if it is to be held longer than about 10 minutes.] Fertilized eggs rotated 15–20 minutes after insemination and fertilization were measured by scoring eggs in morula or late blastula stage 3.5–7 hours after insemination.

B. URODELES

1. Mating

Laboratory-bred axolotls reach sexual maturity at about 1 year of age, unless retarded by insufficient food or improper care. Mature males show a marked enlargement of the lateral margins of the cloacal opening, resulting from the increased size of the cloacal glands as they reach a functional state. The corresponding glands of the female enlarge only slightly, but enough to distinguish a normal female from a castrated or an immature animal. Maturity in the female is usually well indicated by the plumpness of the body resulting from the increased size of the ovaries and oviducts.

The usual breeding season for axolotls under laboratory conditions extends from November or December to the following June. If the animals are maintained between 20 and 22 °C (68–72 °F), spawnings may be obtained even in summer months, especially if animals just reaching the age of 1 year are mated. The cloacal margins of males that have been in breeding condition through the winter and spring are usually reduced in size for several weeks during the summer and fall. During this period the ducti deferentes are devoid of spermatozoa. Similarly, the ovaries of females that have spawned once or more during the spring do not contain mature ova, but only immature growing eggs and degenerating old ones. Such females, if of the white strain, will probably lack the black toe tips noted on animals in good breeding condition.

An aquarium about 0.30×0.46 m (12×18 in.) or a 0.38-m (15-in.)-diameter dishpan are good mating enclosures. The enclosure should contain dechlorinated water that need be no deeper than 0.10–0.15 m (4–6 in.);

the bottom should be covered with a thin layer of fine gravel or very coarse sand. This serves to anchor the spermatophores deposited by the male and to hold them upright; this is essential to permit the female to make proper contact for insemination. Spermatophores do not adhere to smooth glass or enamel and hence are swept aside by the movements of the animals in such enclosures.

When in the proper breeding conditions, no preliminary "conditioning" is necessary to ensure successful axolotl mating. The water in the mating enclosure should, however, be at a temperature no higher than that in the animals' own bowls or aquaria. During the summer months, cooling the water with ice may be advantageous.

When mating axolotls, it is advisable to place a single male and female together in the early evening and leave them undisturbed through the night. The mating enclosure should be darkened by appropriate covering and be placed in a quiet location; bright lights or extraneous sounds may terminate their courtship behavior (Arnold, 1972).

The spermatophores consist of an almost transparent base, are roughly pyramidal in shape, and are constructed of a gel-like material secreted by certain of the cloacal glands. This base is capped by an opaque white mass of spermatozoa that may freely project in cylindrical form. The spermatozoa are held in a compact arrangement, probably by a substance secreted by more internally situated cloacal glands. From one to 25 or more spermatophores may be deposited at a single mating. If none is seen at first glance, stirring the water should reveal them. If the spermatophores lack the apical white mass of spermatozoa, spawning is unlikely to occur. Although males occasionally emit such spermatophores at the beginning of the mating season, they become fertile within a few weeks when spermatozoa have filled the ducti deferentes. If no spermatophores are found, it is advisable not to remate either animal for at least 2-3 days. In any event, return the two animals to their own enclosures instead of leaving them together.

2. Spawning

Ovulation in the axolotl is rarely spontaneous; it ordinarily occurs only after insemination. Eggs are shed into the peritoneal cavity from which they make their way into the cephalic ends of the oviducts. It is probable that the passage of the eggs through the ducts coincides with a period of considerable activity; the swimming of the female becomes somewhat frantic in the final phases of the process. During the spawning itself, the female may remain quiet for considerable periods punctuated with active swimming.

Although low temperature may delay the onset of spawning by several hours, the initiation of spawning is normally expected within a few hours of a successful mating, i.e., within 18-30 h after the animals are placed together. Most eggs are usually laid within 24 h, a factor that can also be delayed by lower temperatures. Sometimes all the eggs are laid during the night following the mating.

The number of eggs per spawning, although varying from a few dozen to several hundred, usually ranges from 300 to 600. The percentage fertility is highly variable and probably depends in large part on the number of spermatozoa received at mating and stored in the glandlike spermathecal tubules in the roof of the cloacal chamber. The stored spermatozoa fertilize the eggs as they pass through the cloaca as revealed by the fact that eggs removed from the oviducts are never fertilized. Fertility may be affected also by the condition of the eggs. When egg overripeness is responsible, this may be recognized by the degenerate appearance of the eggs. If the spawning is at all prolonged, the last eggs laid usually are infertile because they are overripe and incapable of further development.

Occasionally, spawning occurs without insemination, for example, when all spermatophores lack spermatozoa. Sometimes many normal spermatophores may be found after a mating yet all the eggs laid may be infertile; ovulation in such cases must have been induced by the stimulus of the courtship or possibly by a mechanical stimulation of the cloaca.

While spawning, the female may be gently transferred to a fresh enclosure with a clean glass, slate, or enameled bottom. Plants, sticks, or glass rods to which eggs may be attached are quite unnecessary. The eggs will be attached singly or in small groups to the bottom of the container. If not detached by the movements of the female, they may be scraped free with a razor blade or the edge of a glass slide and removed with a pipette of suitable diameter equipped with a large rubber bulb.

Spawning females should be disturbed as little as possible. Cover the enclosure during the day. Unless necessary for experimental purposes, remove the eggs only at intervals of several hours and, if possible, without transferring the female to another enclosure. Some females, if handled or frequently disturbed, may discontinue spawning and retain the eggs, sometimes for several days or even weeks. Occasionally, the swelling of the jelly on a large mass of retained eggs may cause the oviduct to rupture, which will kill the female.

Axolotl females that spawn early in the breeding season may be mated again after an interval of 6 or 8 weeks or sooner if the first spawning was fewer than 100 eggs. A third spawning is frequently obtainable from animals 1-2 years of age. Matings of older females are less successful. It is advisable to plan replacement of breeding animals at the end of their third year, although an occasional one may be useful for a year or two longer.

3. Artificial Insemination

Induced ovulation and artificial insemination of axolotl eggs is possible and useful for certain experiments or when off-spring are desired from individuals that cannot be successfully mated. The procedure is as follows:

Induce ovulation by intramuscular injection of 180–200 IU of follicle-stimulating hormone (FSH). The female is held by wrapping the head and upper body in a wet cloth, and the hormone is injected into the muscle dorsal to the hind legs or cloaca.

At the first evidence of spawning—usually 18–24 h after injection—keep the female *disturbed* for a few hours by occasionally prodding, tilting, or shaking the enclosure. This prevents normal expulsion of the eggs and results in their accumulation in the caudal ends of the oviducts. Do not refrigerate: Not only will refrigeration stop spawning but it probably will also stop oviduct activity. Eggs that are spawned may be fertilized if removed from the water at once and placed on absorbent paper in a covered dish. Without this difficult precaution, spawned eggs exposed to water even briefly are useless for insemination and must be removed surgically as described below.

After eggs have accumulated in the oviducts for several hours, apply gentle pressure to strip any eggs from the ducts. This may result in saving many eggs that might otherwise be damaged in later procedures. The female is then decapitated and pithed. The peritoneal cavity is opened, and blood and remaining eggs in the cavity are rinsed out and discarded. The cut is extended caudally to open the cloacal chamber.

Using two pairs of small forceps, tear open the oviduct, starting at its cloacal end. With the forceps, transfer groups of eggs to clean dry Syracuse watch glasses or small Petri dishes. The eggs, which will adhere to the glass, should be spread out in a single layer. Farther cephalad in the oviduct, the eggs may be spaced singly. Remove each one by grasping the jelly with the forceps, avoiding damage to the egg. Eggs in the slender upper end of the duct that lack membranes and jelly and cannot be removed in this fashion are useless because they will not be fertilized. Avoid smearing eggs with blood.

Cover each dish of eggs as they are collected to avoid drying. Eggs so cared for can be successfully inseminated an hour or more after removal from the ducts.

Decapitate and cut away much of the ventral body wall of the male. Rinse the peritoneal cavity to remove any blood. The ducti deferentes are very slender near the cephalic ends of the testes; when transected here, the ducti do not discharge seminal fluid. Grasp a duct at this region and carefully free it along its entire length by cutting the supporting membrane with fine scissors. Before transecting the duct at its caudal end, grasp it with forceps to prevent the contents from escaping. If there is blood on the duct, rinse it quickly.

Transfer the duct to a watch glass or stender dish containing 10 ml of 10 percent physiological solution. With forceps tear the duct into short pieces to allow the contents to escape. Stir to ensure a uniform suspension, and pipette this over the surface of the eggs, making sure all are moistened. Run a needle or wire several times beneath the sheet of eggs to permit the suspension to pass beneath them as well.

Both ducts of the male may be used to prepare one suspension, or the second duct may be removed some time later if the eggs of a second female are to be inseminated. Merely avoid drying the duct if it is not being taken out immediately. If removal of the eggs has proved quite time-consuming and only one female is being used, it may be advantageous to remove and inseminate the eggs from one oviduct before proceeding with removal of the eggs from the second. If two females are being used, a second person should assist, although both ova and spermatozoa will survive considerable delay without apparent injury.

Inseminated eggs should be allowed to stand for about 20 min and are then flooded with medium. After another 20 or 30 min immerse the small dishes containing eggs in a large bowl of medium. Later, the sheet of eggs may be detached from the watch glass with a razor blade or scalpel. By using two needles, somewhat like the blades of scissors, the sheet of eggs may be subdivided without breaking the egg membranes. When the eggs are in early blastula stages, remove those that are infertile or broken, using the needles to detach them from the others.

4. Initial Care of Embryos

Eggs obtained either by artificial insemination or from natural spawning require chlorine-free water. No more than 50-60 eggs should be cultured per liter medium in a shallow bowl in order to facilitate gas exchange. At 20 °C (68 °F) cleavage will begin 6-8 h after spawning or insemination; sorting, however, should not be attempted until early blastula stages several hours later. Failure to remove infertile or damaged eggs may contaminate the healthy embryos.

Hatching occurs at about 2 weeks, but may be delayed by keeping the eggs at 10-15 °C (50-59 °F). Temperatures lower than 10 °C (50 °F) are inadvisable, and even that temperature may be harmful if exposure is prolonged. The care of later stages has been described in Chapter VI, Section C.1.

VIII Records and Information Control

A. IDENTIFICATION OF INDIVIDUALS

1. Numbering

The identification of individual animals—often necessary for the management of experimental animals—is mandatory in breeding and genetic procedures. Several techniques can be used to identify individual amphibians. Selection depends on the species and the developmental stage. Ideally, the system of choice would allow identification of individual animals at all stages of development from the time of fertilization through larval and postmetamorphic development. However, aside from isolation in individual enclosures, no system adequate to accomplish this objective is available. In practice, a clutch of eggs or a shipment of animals may be given a serial identification number such as 35,000 with subsequent numbers in the series reserved in proportion to the number of clutch or shipment members to which individual numbers will be assigned. As animals are assigned for specific experimental purposes, the experimental group or individual may be given numbers from among those reserved. Thus, 35,XXX identifies the clutch or shipment; the hundred's digit may be used to identify groups and the ten's and unit's digits to identify individuals. Serial numbers that do not include letters are recommended to facilitate adaptation to computer techniques, should these be found useful (see Section B below). Identification numbers may be written in waterproof ink on waterproof labels that, when appropriate, may be easily transferred from one enclosure to another. In the Amphibian Facility of the University of Michigan, it has been found useful to assign unique group numbers to members of a clutch or shipment that share an enclosure. This permits tracing enclosure mates and identifying individuals with common environmental experiences.

2. Tattooing

Tatooing by ink injection techniques has proved successful. It is particularly useful for larvae with well-developed tail fins, although its use is not limited to them. In this technique the animal is lightly anesthesized (see Chapter IX, Section F) and a fine needle attached to a hypodermic syringe containing tattoo ink is inserted into the tail fin or into the skin. The ink is injected as the needle is withdrawn, leaving a "bar" of ink. By using inks of several colors and forming a series of parallel bars, unique identification codes may be constructed. Recently, several useful variations on this principle have been published (Woolley, 1973).

Tattooing by the use of vibrating needles, as used in human tattooing, cannot be applied to larvae or small postmetamorphic animals; in fact, this method can only be used for larger postmetamorphic animals and even then must be renewed frequently. *R. pipiens* and bullfrogs retain such a tattoo for only 3-4 months. Because these and other amphibians shed their skin, it is difficult to assure that the needle-driven tattoo penetrates into skin layers that are not lost.

3. Branding

Wolf and Hedrick (1971) describe "chemical branding" as a procedure for marking *Xenopus*; these are permanent for at least a year. Using a cotton swab dipped in a solution of 0.5 percent amido Schwartz in 7 percent acetic acid, label figures are formed on the back of an animal after mucous secretions are removed by repeated wiping with paper tissues. After 1 min of contact the animals are returned to an enclosure. The dye provides a temporary identification until scar tissue is formed a few days later in response to the acid treatment.

The cold branding technique (Farrell and Johnson, 1973) has proved satisfactory for adult *R. pipiens* and *R. catesbeiana* and may be applied to the dorsal surface where the label can be easily read. The advantage of this technique over tattooing is the longer period of label retention. However, cold branding also disappears with time and cannot be applied to juvenile animals. Tattooing and cold branding are preferable to heat branding because they result in less trauma to the subject.

Urodeles can be tattooed or branded at relatively early stages if it is done under light anesthesia (see Chapter IX, Section F). Such markings are lost, however, as the skin is shed and must be renewed periodically.

4. Toe Clipping

Toe clipping is an effective technique for amphibian identification and is appropriate for most anurans that do not readily regenerate lost digits. However, it is inadequate for urodeles and *Xenopus* in which the regeneration process replaces the removed digits, unless the regeneration is inhibited by treatment with berilium nitrate (Heatwole, 1961).

Ranidae may be toe clipped shortly after metamorphosis using light anesthesia and cuticle scissors. Although older *R. pipiens* may be toe clipped with scissors, mature *R. catesbeiana*, because of their size, may bleed excessively and danger of local infection occurs when toes are clipped in this manner. However, toe clipping large anurans may be accomplished by using an appropriate cautery knife or loop to seal the wound at the time of amputation. A coding system for toe clipping based on the system for punching edge coded cards and that allows 9,999 individuals to be distinguished with only two amputations per foot is illustrated in Figure 24.

5. Other Marking Systems

Various systems for labeling amphibians by the insertion of plastic rings and other devices have been attempted but without notable success.

FIGURE 24 Toe clip conventions (Nace *et al.*, 1973; reproduced with permission from the *American Zoologist*).

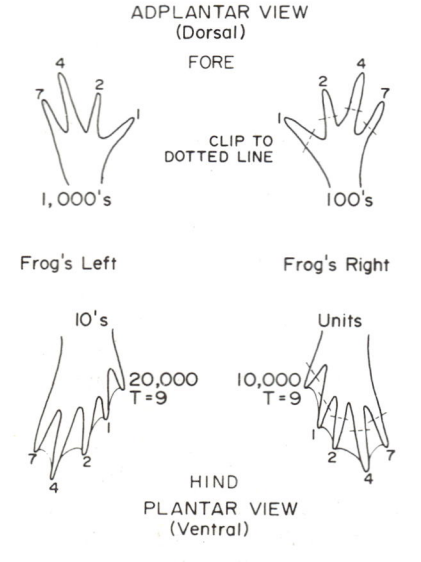

30,000 = BOTH 20,000 AND 10,000 CLIPPED ABOVE 39,999, TREAT AS FROM 1 TO 39,999

6. Drawings and Photographs

The most foolproof method of identifying individual amphibians is the use of photographs that record the details of natural disruptive patterning. Figures 25, 26, and 27 illustrate disruptive patterning in *R. pipiens* and how these patterns are coded for identification purposes in the Amphibian Facility of the University of Michigan (Nace *et al.*, 1973). Similar classification of the patterns of other amphibians is possible, but has not as yet been completed.

In laboratories with small numbers of animals, it is adequate to prepare duplicated outline drawings of the amphibian and to fill in these outlines with drawings of the disruptive patterning. This technique, however, is too time-consuming for a large colony of animals; thus, photographs are recommended. Using the classification system, it is possible to identify quickly the class of patterns to which an animal belongs and then, by comparison with the appropriate specific photographs or drawings, to identify the individual animal. When the pattern classification and recorded pattern are used in conjunction with other characteristics—such as sex, last recorded snout-vent length, and other unique characteristics—a very large number of individuals may be identified.

B. INFORMATION CONTROL SYSTEMS

The management of animal colonies or of data collected in even small colonies becomes increasingly complex as the numbers of animals or the history of the colony increases. This problem can be greatly alleviated by the adoption of computer-based techniques. Recent improvements in available computer programs and time-sharing techniques, even over long-distance telephonic connections, permit even those inexperienced in computer use to adopt these procedures readily. Although they may be adopted at anytime in the history of a colony, the earlier they are used, the greater the economy that is realized. Current technology is sufficiently advanced that standardization of the data base or of the software is not necessary. It is sufficient to record the available data in machine-readable form; the computer itself can restructure the data in accordance with the requirements of specific current or future software.

A computer-based system currently in use for the management of amphibians is described in some detail in Nace *et al.* (1973). This system is in the public domain and accessible to long-distance users. Other systems that may be adaptable to users of amphibians include the LRE system (Laboratory Research Enterprise, Inc., Kalamazoo, Mich.) for Beagles and a system used for microorganisms (Bachmann *et al.*, 1973).

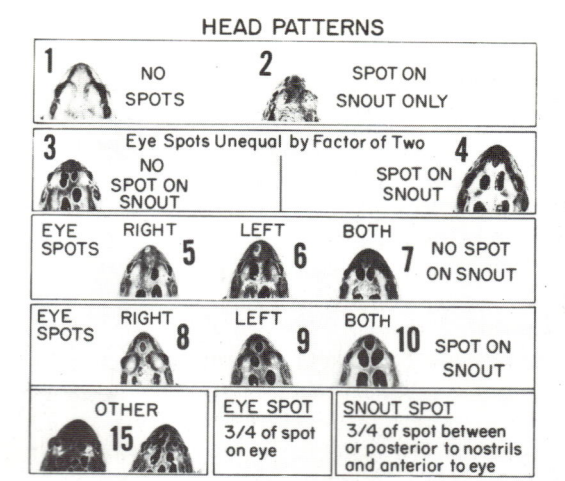

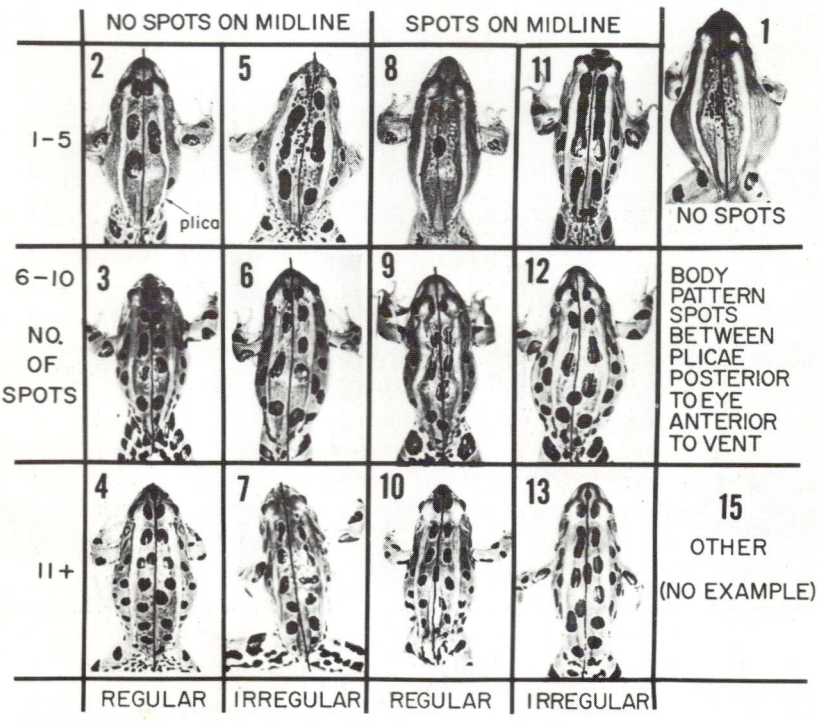

FIGURE 25 *Rana pipiens* head and body pattern code characteristics. The numbers appearing by each pattern in Figures 25 and 26 are used to write an identification formula. Thus, 7-6-2-4 identifies the animal shown with body pattern "6" (Nace *et al.*, 1973; reproduced with permission from the *American Zoologist*).

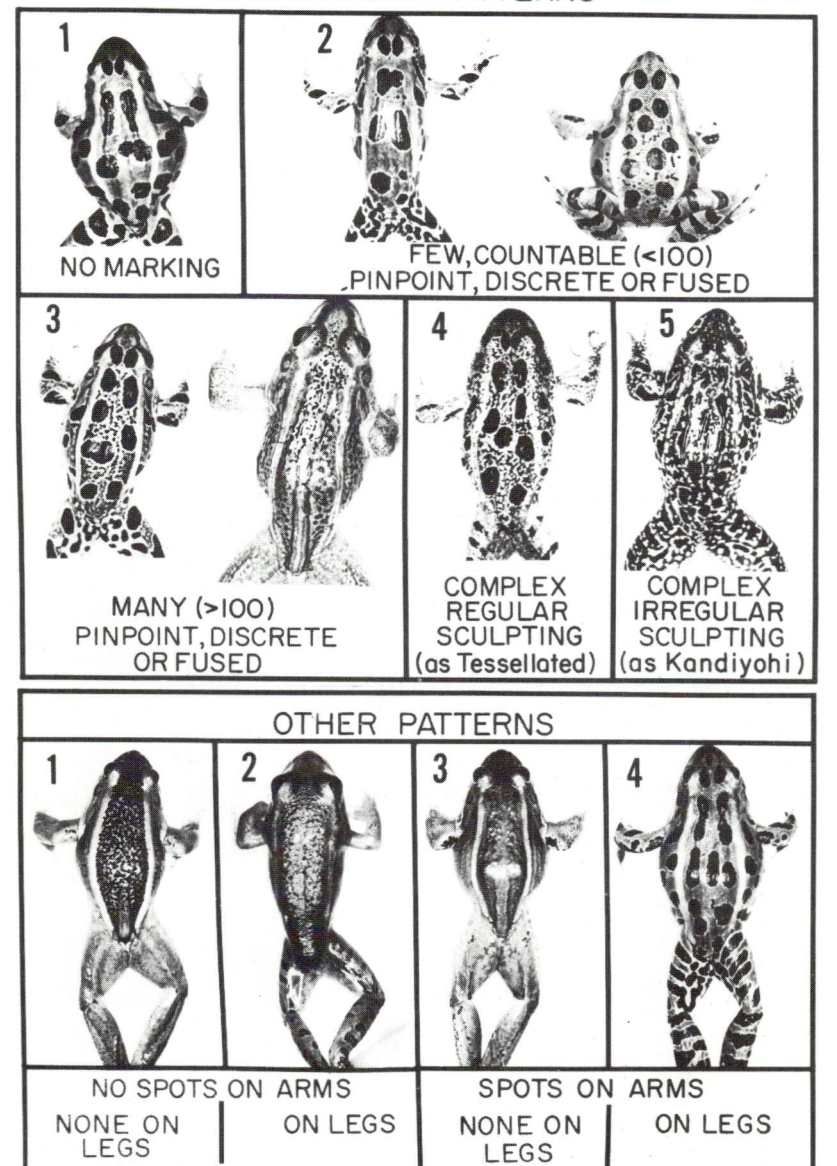

FIGURE 26 Background and other patterns (Nace *et al.,* 1973; reproduced with permission from the *American Zoologist*).

SPOT:
DETERMINED BY CHARACTER OF SURROUNDING SPOTS

FUSED SPOT
= ONE SPOT
= IRREGULAR

SPOT ON MIDLINE:
WHEN 1/3 OR MORE OF SPOT OVER MIDLINE

REGULAR: APPROXIMATES A SQUARE, OVAL OR CIRCLE

IRREGULAR:
NOT SYMMETRICAL, INDENTED OR LONG

LONG SPOT:
3 x (or more) AS LONG AS AVERAGE WIDTH

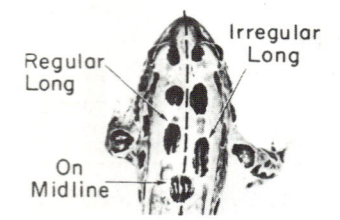

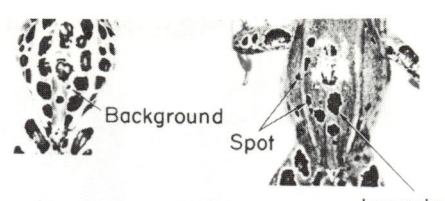

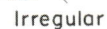

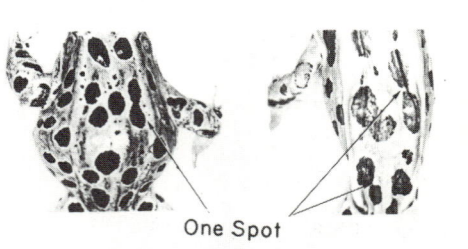

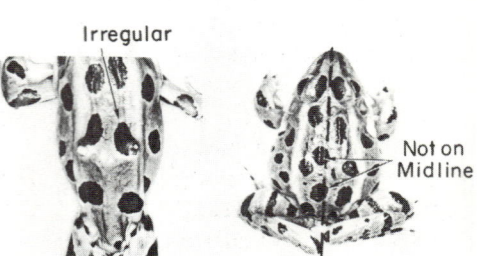

FIGURE 27 Conventions used to characterize spots (Nace *et al.*, 1973; reproduced with permission from the *American Zoologist*).

IX Amphibian Medicine

A. GENERAL COMMENTS

Considering the wealth of information dealing with the physiology, biochemistry, genetics, developmental biology, and even behavior of amphibians, it is incredible that almost nothing of a practical nature is known about amphibian medicine. Entire books deal with the classification of amphibian infections and infestations, while therapy and prevention are usually dispensed with within a few pages. Moreover, those methods suggested, for the most part, have never been adequately evaluated. Yet, many amphibians brought to the laboratory do not survive for more than a few days under conditions effective for maintaining "healthy" wild-caught and laboratory-reared or laboratory-bred animals for years. No doubt our inability to define or measure amphibian disease in anything but the crudest terms has contributed to this situation. An amphibian can have a "good" appearance and yet be within hours of death as the result of massive destruction of internal tissues. On the other hand, we easily confuse mild inflammation of the skin with serious disease. Certainly, anyone using amphibians in their research should evaluate the potential impact of "disease" on their investigations and, insofar as possible, define the state of health of the individuals being studied.

While some diseases of amphibians interfere with laboratory experiments, other diseases may be exploited as animal models of human disease. Indeed, much of the interest in the renal adenocarcinoma of *R. pipiens* is not motivated by interest in the hazard of the tumor to frog populations but is related to the similarity of the frog cancer to comparable tumors in humans (Duryee *et al.*, 1960; Dawe, 1969).

It is beyond the scope of this document to review all the diseases of amphibians and all of the suggested treatments (see Boterenbrood, 1966; Frazer,

1966). Since frog medicine is a relatively advanced specialty of amphibian medicine, it will serve as the best model and the following discussion will be restricted to several common frog diseases and how they may affect *R. catesbeiana* and *R. pipiens* in the laboratory. Additional background information and discussions regarding amphibian diseases may be found in Elkan (1960), Gibbs (1963, 1973), Walton (1964, 1966, 1967), Inoue *et al.* (1965), Reichenbach-Klinke and Elkan (1965), Gibbs *et al.* (1966), Joiner and Abrams (1967), Abrams (1969), Crans (1969), Lom (1969), Mizell (1969), Rowlands (1969), Boyer *et al.* (1971), Mawdesley-Thomas (1972), Van der Steen *et al.* (1972), Cicmanec *et al.* (1973), and van der Waaij (in press). Serious students of amphibian diseases may also find the extensive collection of reprints at the Osborn Laboratories of Marine Sciences to be an invaluable source (Ross R. Nigrelli, Chief Pathologist, Osborn Laboratories of Marine Sciences, New York, New York).

B. BACTERIAL DISEASES

1. Pathogens

Bacteria and viruses deserve special attention because they are responsible for the majority of deaths that occur in populations of laboratory frogs. Also, in the case of bacterial infections at least, there are methods that seem to be effective in preventing and treating the diseases. Although the list of bacteria identified from amphibian sources is lengthy (Walton, 1964, 1966, 1967; Reichenbach-Klinke and Elkan, 1965), only a few seem to be major pathogens. Thus, despite the complexity of the bacterial millieux, only *Aeromonas hydrophila* (Ewing *et al.*, 1961) has been repeatedly implicated so far in the large-scale mortality of leopard frogs since 1898 when Russel first isolated an organism that was probably *A. hydrophila* and that he called *Bacillus hydrophilus fuscus* (Russel, 1898). *A. hydrophila* has also been called *Proteus hydrophilus* and *Pseudomonas hydrophilus* (Gibbs *et al.*, 1966; Reichenbach-Klinke and Elkan, 1965). The disease produced by *A. hydrophila* has been "red-leg" but the symptomatology is not sufficiently consistent or specific to warrant this title. Thus, Miles (1950) reported an outbreak of "red-leg" in tree frogs at the London Zoo but identified *Bacterium alkaligenes* as the causative agent. To avoid confusion, it would be better to define diseases on the basis of the bacteria responsible.

An essential step in the study of any disease is the isolation and classification of organisms from sick animals. In a major study of northern *R. pipiens* exhibiting a wide variety of symptoms, three organisms were found to be associated with serious illness (Gibbs *et al.*, 1966). This study is noteworthy because it demonstrates the effectiveness of a systematic methodology for studying pathogenic organisms and the treatment of the diseases they produce. Such methods have not been widely practiced in amphibian medicine. Blood samples for bacterial culture were obtained aseptically. Cultures were incubated at 37 °C (98.6 °F) and, in order not to overlook psychrophilic bacteria, at 25 °C (77 °F) for 1 week before any were assumed to be negative. The samples were discarded if the animals from which they were obtained failed to survive for at least 8 hours. This precaution reduced the possibility of isolating nonpathogenic bacteria that might enter the blood and other tissues under conditions of shock and widespread collapse of tissue functions. *A. hydrophila* and types of *Mimeae* (Ballard *et al.*, 1964) associated with *A. hydrophila* were found. The *Mimeae* are easily missed as they tend to grow relatively slowly and the *Aeromonas* overgrows them. These observations led to the conclusion that *A. hydrophila* and *Mimeae* were the major pathogens found in populations of *R. pipiens* from the north central United States (Gibbs *et al.*, 1966). Although *Staphylococcus epidermis* was isolated repeatedly from purulent leg infections, such infections were not common and the causative role of *Staphylococcus* remained unclear. Doubtless, many more organisms will eventually be identified as etiological agents of amphibian disease once techniques and observation become more refined.

2. Drug Selection and Administration

The use of antibiotics should be guided by determining the antibiotic sensitivities of the organisms in question. This is readily done by using standard tests in which effective inhibition of growth serves as the indicator of sensitivity to the antibiotic. This method was applied to each of the three organisms isolated in the study referred to above (Gibbs *et al.*, 1966) and Tetracycline · HCl was selected as the drug of choice on the basis of sensitivity and wide tissue distribution observed in other animals.

Other infections may require other drugs, few of which have been adequately tested on frogs. Since so little is known about the sensitivity of amphibians to drugs or about the fate of administered drugs, the potential complications of antibiotic therapy, both to the health of the animal and to the results of the experiments in which the animal is to be used, must be thoroughly researched before using a drug routinely.

The Tetracycline treatment of *R. pipiens* provides a glimpse of possible complications (Gibbs, 1963; Gibbs, *et al.*, 1966). It was determined that placing Tetracycline in the tank water serves no useful purpose. It is not ingested in sufficient amounts or absorbed sufficiently to result in significant blood levels. Attempting to raise the water concentration of the drug only results in deaths due to skin damage. Intraperitoneal injection of Tetracycline is also contraindicated because of local irritation, although attempts to ameliorate this have been reported. The only route tested that has been generally effective in *R. pipiens* is the oral route. (Other routes may be possible with other drugs or other amphibians.) In addition, Tetracycline must be administered by stomach tube since the drug is very bitter and stimulates a gag or vomiting response.

Stomach tubes (Figure 28) are conveniently made from polyethylene uretal catheters or other small-gauge soft polyethylene tubing following the procedure described by Gibbs (1963). The essential feature of the stomach tube is a small ball tip, which protects tissues from damage. The tube is open just behind the ball tip. Plastic disposable syringes with plastic plunger seals are recommended for use with the stomach tubes since the plungers do not slide while the stomach tube is being inserted. The stomach tube should be inserted gently into the frog's esophagus with light manual pressure so that the ball tip passes the pyloric sphyncter without unnecessary irritation and vomiting. Stomach tubes should be maintained in good repair, free of rough surfaces, to further minimize irritation. Smooth-tipped

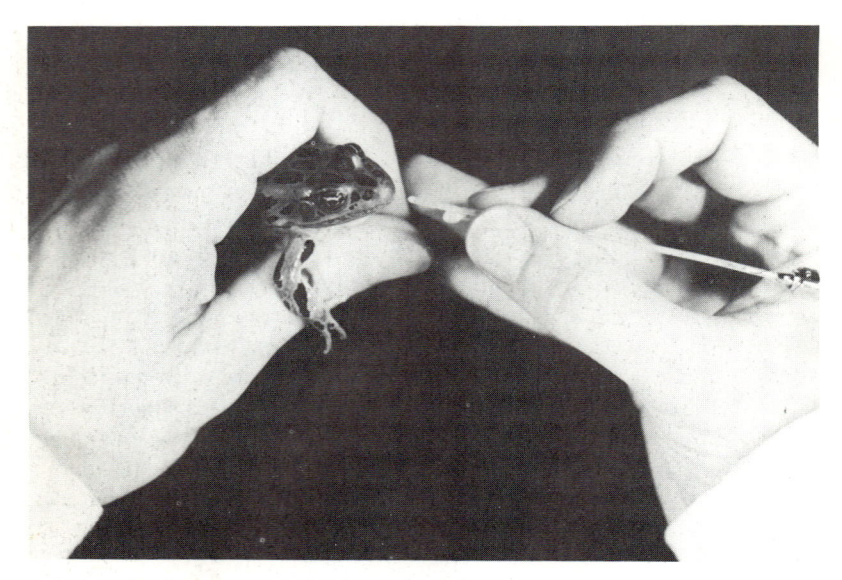

FIGURE 28 Use of a stomach tube on *R. pipiens*. Note that the length of the tube allows gentle, fingertip manipulation while the syringe rests on the base of the hand. The second blunt instrument seen below the stomach tube is inserted between the jaws of the frog to stimulate opening of the mouth and avoid injury by the larger stomach tube.

eye droppers have been used successfully with *R. catesbeiana,* and fabricated stomach tubes for small animal work are available commercially.

Soluble Tetracycline · HCl should be made up at known concentrations of about 25 mg/ml in distilled water so that the injection volume can be kept to approximately 0.2 ml for *R. pipiens,* as larger quantities are easily regurgitated. Approximately 1 ml may be given to *R. catesbeiana* without a serious risk of regurgitation. A number of small aliquots of the solution should be prepared to minimize the chance for contamination, which may cause the Tetracycline to precipitate. The stomach tube and syringe should be rinsed with distilled water after each use and periodically with dilute hydrochloric acid in order to prevent precipitation within these parts. During treatment, the frogs should be kept in sloping tanks, half filled with water. *Care should be taken to keep the water clean since, as mentioned above, the Tetracycline can damage the frog's skin and much of the dosage is excreted or passed unchanged.* Food is not well digested and therefore not offered during treatment, but it is very important that the frogs be maintained in an optimum environment with optimum temperatures (see Chapter VI) as their immunity systems are very temperature dependent (Volpe, 1971).

Twice daily administration of a dosage of 5 mg Tetracycline · HCl/30 g body weight produces effective blood levels and is well tolerated. Although apparently unnecessary, the dosage can be doubled without ill effects. The effectiveness of the dosage was tested *in vivo* by injecting a control and a treatment group of frogs with a live culture of *Aeromonas hydrophila* (Gibbs, 1963). The treatment group was placed immediately on a 1-week course of Tetracycline. Of the untreated group all died within 9 days and of the treated group only one died on the first day and none thereafter. Similar experiments with varying periods of treatment have demonstrated that a 5- to 7-day course of therapy is necessary in most cases. The usefulness of the Tetracycline treatment has been confirmed in many laboratories (Papermaster and Gralla, 1973).

Once the frogs have completed a 5- to 7-day course of treatment, they should be offered live food and an optimal environment (see Chapter VI) for a recovery period of at least 3 weeks. This period allows diseased or damaged tissues to heal and most residual Tetracycline to be excreted. Note, however, that some Tetracycline will be bound to tissues containing calcium and may be retained for long periods of time. Although wide spectrum light is recommended for most amphibians (see Chapter V, Section B.2.c), frogs treated with Tetracycline should not be exposed to ultraviolet light for about 2 weeks after the termination of medication since this drug may produce photosensitivity.

The remarkable recovery demonstrated by treated frogs attests to the effectiveness of their immunity system since Tetracycline is not bactericidal under the conditions of the treatment. This fact has important implications regarding the contagion of bacterial disease and the need for isolation procedures and sterilization of tanks. Tank sterilization, as distinct from cleanliness, is not necessary; *A. hydrophila* can be readily isolated from the feces of healthy, treated frogs. Consequently, if septecemia attributable to *A. hydrophila* develops, it must represent a response to some factor that triggers a decrease in the frog's bacterial defenses. This factor might be viral, nutritional, other as yet unidentified bacteria, or stress. At present, the relative significance of these possible factors cannot be evaluated, but the user is cautioned to be sensitive to their possible role in the disease process. From a practical standpoint, Tetracycline-treated frogs have been held for years without recurrence of serious bacterial infections (Gibbs *et al.*, 1966). The major cause of death in such a treated population is likely to be the Lucké renal adenocarcinoma.

3. Identifying Diseased Frogs

Perhaps the most common mistake made by those working with frogs is to attempt to distinguish between healthy and sick animals purely on the basis of appearance. Nearly half of the frogs heavily infected with *A. hydrophila* will exhibit only mild somatic signs, such as being thin or lacking brilliance of the skin coloration. Their behavior may lack purpose and they may be either hyperactive or sluggish. Since these signs overlap with those of uncomplicated malnutrition, fright, cold torpor, or the irritability associated with overheating, it is quite impossible to identify healthy frogs with certainty. Shortly before death, frogs, including some which may even have looked "healthy," usually vomit blood and convulse. When present, the more striking and traditionally recognized symptoms include slumped posture (palms turned outward), disinclination to move when prodded, tense abdomen, cutaneous hemorrhages, eroded toes and feet with bare bones exposed, eroded jaws, perforations of the skin on the dorsal surfaces (particularly the nose), rough and bleeding nictitating membranes, hemorrhaging within the eyes, and numerous neurological signs. The course of illness may extend over several months. Spontaneous recoveries are rare, although several authors have reported them (Russel, 1898; Emerson and Norris, 1905). The presence of *Mimeae* in conjunction with *A. hydrophila* does not result in specific symptoms or special complications differing from those already described (Gibbs *et al.*, 1966).

4. The Need for Treatment

Since bacterial disease can seriously affect experimental results, treatment or prevention is essential. This is emphasized by the extensive histological studies of animals infected by *Aeromonas*, which date from Russel (1898). The work of Russel and later investigators (Emerson and Norris, 1905; Kulp and Borden, 1942; Rose, 1946; Gibbs *et al.*, 1966) has shown that virtually every organ and tissue of the frog is affected by this organism.

As one illustration of the physiological impact of such infection, a recent study of the sartorius muscle of infected *R. pipiens* is instructive. The muscles of newly arrived untreated and treated and recovered frogs were examined histologically and physiologically. The untreated frogs were carefully selected to exclude certain diseases but to include *Aeromonas* and *Mimeae* infections and malnutrition, which could not be studied separately. The muscles from diseased animals showed a combination of depleted metabolic stores, edema, necrosis, and hemorrhage that would introduce serious errors in biochemical determinations made on such muscles. Function, as measured electrophysiologically, was also seriously impaired in the muscles of the untreated animals (Gibbs, 1973). Unquestionably, use of the frog as a model of physiological processes suffers the same limitations of all animal model research. If the frog is not healthy, the value of the investigation must be held in serious doubt.

C. VIRAL DISEASES

1. General Comments

As a practical matter, it is virtually impossible to obtain virus-free frogs, yet very little can be done to treat virus infections (Lunger, 1966; Mizell, 1969; Granoff, 1969, 1972). Consequently, the investigator must study animals infected with viruses that are, for the most part, unidentified, of unknown effect on the host, and of equally unknown effect on the experiment (Whipple, 1965). Since no effective, specific treatments are available for any viruses, the only protection for an amphibian colony is good animal husbandry.

As already mentioned, the Lucké renal adenocarcinoma virus or Lucké tumor herpesvirus (LTHV) is the major cause of death in populations of laboratory frogs that have been treated for their bacterial infections and maintained under optimal conditions. In the experience of one laboratory, it appears that, provided the treated frogs are held long enough, almost all of them will eventually succumb as the result of kidney tumors. Apparently, the virus or its oncogenic effects can remain latent for years. Frogs held at warm temperatures do not shed the Lucké virus, but refrigerated frogs develop the virus quickly in the kidneys. Tadpole edema virus may also be responsible for the loss of adult laboratory frogs. It is felt that occasional epidemics of bloating, eventually leading to death, may be related to this virus, which is highly lethal in tadpole populations. For the most part, however, adult frogs appear to be immune to the virus (Wolf et al., 1969). Kidney tumors do not have an obvious effect on the health of frogs until the tumors destroy essentially all useful kidney tissue or until they metastasize to other more critical tissues, such as the lungs or liver.

2. Lucké Tumor Herpesvirus (LTHV)

The virus of the Lucké renal adenocarcinoma is believed to be a herpesvirus; it is icosahedral with a capsid consisting of 162 capsomeres (Lunger, 1964) and has a DNA core (Zambernard and Vatter, 1966; Wagner et al., 1970). A substantial body of evidence has emerged to suggest a causal relationship between LTHV and the Lucké renal tumor (McKinnell, 1973). Tadpoles injected with virus-containing extracts develop tumors (Tweedell, 1967, 1972; Mizell et al., 1969), and LTHV is omnipresent in all cold weather renal tumors (McKinnell and Ellis, 1972). The viral genome is present in summer-phase "virus-free" tumors (Collard et al., 1973). Tumor prevalence varies from 6–9 percent for grossly observable neoplasms (McKinnell, 1965) to almost 100 percent for tumors detectable by microscopic examination (Marlow and Mizell, 1972). Frogs with this herpesvirus-associated tumor will ultimately succumb to the disease in the laboratory.

Another herpesvirus, known as frog virus 4 (FV-4), was isolated originally from the urine of a tumor-bearing frog (Rafferty, 1965). The DNA of FV-4 differs from that of LTHV (Gravell, 1971), and the virus does not produce tumors when injected into tadpoles. Its effect on the health of the frog is unknown.

3. Amphibian Polyhedral Cytoplasmic Deoxyribovirus (PCDV)

This virus—also known as frog virus 3 (FV-3) (Lunger, 1966) and tadpole edema virus (TEV) (Wolf et al., 1968)—was isolated from normal and tumorous *R. pipiens* (Granoff et al., 1965) and from *R. catesbeiana* larvae of a number of localities in the eastern United States (Wolf et al., 1969). Injections of FV-3 are lethal to embryos and larvae of *R. pipiens* (Tweedell and Granoff, 1968) and will cause death of young *R. catesbeiana* when added to aquarium water (Wolf *et al.,* 1969). Morphology as revealed by electron microscopy and biochemical studies suggests that amphibian polyhedral cytoplasmic viruses isolated from leopard frogs, bullfrogs, and newts are indistinguishable (Clark *et al.,* 1969).

4. Lymphosarcoma Virus of *Xenopus*

The South African clawed frog, *Xenopus laevis,* is susceptible to lymphosarcomas (*Xenopus* L-1 and L-2) thought to be caused by viruses (Balls and Ruben, 1967, 1968) that have thus far not been detected with the electron microscope (Hadji-Azimi and Fischberg, 1972).

D. PARASITIC DISEASES

That frogs and other amphibians can be heavily infested with a seemingly endless number of metazoan parasites is well known (Reichenbach-Klinke and Elkan, 1965; Walton, 1964, 1966, 1967), and the list of even the most commonly encountered parasites is too long to recount. In spite of their ubiquity, the effect of parasites on the frog's health is far from clear. Parasites have been blamed as the direct cause of death and indirectly as vectors of virus and bacterial disease. However, parasite infestations tend to decrease in wild-caught frogs as they are held in optimal conditions (see Chapters V and VI) free of intermediate hosts; for all practical purposes, parasites do not appear to seriously affect the health of the laboratory frog (Gibbs *et al.,* 1966). This is not to say that the presence of parasites will not affect the frog's health to some degree, for they can most certainly cause local destruction and irritation of tissues.

Parasitic infestation of frogs, however, must have important consequences for the investigator. The modified physiology resulting from the infestation certainly has a potential for modifying the results of many types of physiological investigations. Biochemical investigations may also be effected by the presence of parasites.

It is because of the implications of parasitic infestations to the investigator that the definitions of experimental animals for the laboratory (Chapter III, Section B) sharply distinguish between those animals directly or indirectly exposed to intermediate hosts and those protected from such exposure. In this regard, the selection of food items is restricted by the fact that living or unprocessed foods may introduce parasites into the laboratory environment unless these foods are, themselves, isolated from the life-cycle sequences of potential parasites. This is illustrated by *R. catesbeiana* deaths experienced at the Louisiana State University amphibian facility when heavy infestations of a nematode, *Eustrongylides* sp., were traced to the use of live fish from outdoor ponds in the diet for the frogs.

Since very little is known about the treatment of the parasitic diseases of amphibians, the only protection available to the amphibians and to the integrity of the experimental results is to maintain the animals in an optimal environment, free of intermediate hosts, or to use animals that meet the requirements defined for the laboratory-bred classification (see Chapter III, Section B).

Lists of parasites that infest amphibians are not given because no recommendations for specific treatments can be made. Those interested in the identification of amphibian parasites are referred to the authors cited above.

E. MYCOTIC DISEASES

Only a relatively few amphibian fungal infections have been identified. *Saprolegnia* is reported to be the most common fungus afflicting amphibia (Reichenbach-Klinke and Elkan, 1965; Walton, 1964, 1966, 1967). It often attacks nonviable amphibian eggs in an egg mass and may pose a threat to the other eggs. *Basidiobolus ranarum* occasionally infects amphibians. Although *Fonsecaea pedrosoi* has been identified in *B. marinus, R. catesbeiana,* and *R. pipiens,* transmission studies using *B. marinus* resulted in infestation and death in only the stressed group; the unstressed group remained free of the disease (Cicmanec *et al.,* 1973). Thus, while fungi can be the direct cause of disease and death in some cases, as a rule, their infestations are secondary to other infections or nutritional disorders. As in the case of parasites, fungi and algae do not appear to represent a primary threat of serious disease in populations of well-maintained laboratory amphibians.

Some authors have recommended increased salinity, dips of potassium permanganate or formaldehyde, or topically applied tincture of iodine as treatment for fungal disease. However, in view of the highly sensitive and vital nature of amphibian skin, such practices must remain open to question, since they may only serve to produce additional insult and the efficacy of such treatments has not been clearly established under controlled conditions.

F. EUTHANASIA AND ANESTHESIA

Considerations of humane treatment and the research requirement that animals in both acute and chronic protocols be subject to a minimum of stress introduce the need to apply the best available procedures for euthanasia and anesthesia. Kaplan (1969) recommends an excessive dose of ether or pentobarbital for euthanasia of all amphibians. The action of these agents is not instantaneous, however, and the level of stress produced is unknown. The least cumbersome and least stress-producing procedure may still be brain and spinal pithing, an instantaneous technique. A sharp needle of a diameter appropriate to the size of the animal is quickly inserted in a cranial direction through the foramen magnum and is rotated in a manner to crush the brain bilaterally. The needle is then inserted in a caudal direction at the same point of entry and rotated to destroy the spinal cord. The location of the foramen magnum may be readily identified by a slight depression in the skin on the midline just posterior to the eyes when the animal is held between the thumb and last three fingers and when the index finger, placed on the head, is used to press the head in a ventral direction. If not immediately evident visually, the tip of the needle may be placed on the skull and slid caudally along the midline. The depression marking the location of the foramen magnum is readily detected as the needle slides over the junction between the skull and first vertebra. For procedures that require that the brain and spinal cord remain intact, we recommend the initial induction of deep anesthesia.

The most useful but most expensive anesthetic reagent is tricaine methanesulfonate (MS-222). Larvae may be fully immobilized within 80 s at room temperature by immersion in a 1:2,000 or 1:3,000 solution of this reagent. Recovery occurs within 4-14 min upon return to normal medium, depending on the stage of development and length of exposure to the reagent. Exposure to MS-222 for as long as 1 h does not result in abnormal development (Kaplan, 1969). Adults may be anesthetized by immersion in the same preparation. This is slow, however, and more certain control of the depth of anesthesia is possible by injecting the reagent into either the dorsal lymph sac or intraperitoneally. Good results are obtained in *R. pipiens* by intraperitoneal injection of 0.1 ml of 1 percent MS-222/10 g of frog. If a frog is not completely anesthetized in 5 min, inject half the original dose. The frog will be in deep anesthesia appropriate for surgical manipulation for about half an hour (Nace and Richards, 1972b). Repeated doses can be used for longer periods, and recovery is good when the animals are set aside on a moist towel in a vegetable crisper. Note that the animal will drown if placed in deep water.

Kaplan (1969) discusses other procedures and also notes that, although there is no literature on the errors introduced into experiments because of the lack of or misuse of preanesthetics, analgesics, or anesthetics, postanesthetic care of amphibians has no problems and requires no special procedures. Among useful procedures Kaplan (1969) describes the use of hypothermia and chloretone (immersion in 0.2 percent solution) and discusses the disadvantages of several other anesthetics that have been used. The investigator who expects to make extensive use of anesthesia in amphibians should consult this paper.

It must be cautioned that anesthesia must be used with care in certain physiological studies. Farrar (1972) demonstrated that repeated treatment of *R. pipiens* with Finquel (MS-222) causes increased blood glucose and lactate and discussed the mechanism of this response.

X Personnel

A. GENERAL COMMENTS

The personnel requirements for a facility maintaining amphibians are not essentially different from those for other laboratory animal facilities. Therefore, pages 15-17 of the *Guide for the Care and Use of Laboratory Animals* (Committee on Revision of the Guide for Laboratory Animal Facilities and Care, 1972) should be consulted. However, insofar as we know, except for The University of Michigan, no other institution routinely trains veterinarians in the care and management of amphibians as laboratory animals. This training results from collaboration between the Unit for Laboratory Medicine and the Amphibian Facility. Consequently, it is doubtful that the recommendations for personnel as given in the above *Guide* and calling for direction by veterinarians can be met at present. Until this situation changes it seems more realistic that direction of an amphibian facility should reside with a staff member experienced in amphibian biology and reproduction and who is advised by a veterinarian experienced in laboratory animal medicine. The former assures that the real differences between amphibian husbandry and the husbandry of other laboratory animals are clearly recognized; the latter assures that well-known and standardized practices that are applicable are followed.

Similar circumstances apply with respect to animal technologists. Currently, no formal training or certification is available for animal technologists wishing to specialize in amphibian care. Although arrangements can be made for technical personnel to gain experience with amphibians at the Amphibian Facility of The University of Michigan, the availability of personnel trained there will remain limited for the immediate future. In the absence of specific training programs it has been found that, for both academic and commercial establishments concerned with the handling of amphibians, certain criteria are of value in selecting personnel. Among these criteria are adaptability, sensitivity to animals as frequently revealed by those with animal hobby interests, and an understanding that work schedules require the assignment of duties on a 24-h-a-day, 7-days-a-week basis. The need for insect as well as amphibian husbandry requires regular attention on weekends. In the absence of training that inculcates the concept of this requirement, weekend duties too frequently must be met by the operating managers.

Other aspects of personnel management are not unique but are of major importance and managers of facilities would be well advised to develop expertise as now understood by specialists in this area (D'Ver, 1973).

B. SAFETY HAZARDS

1. Physical

Among the relatively few hazards in an amphibian facility, the greatest danger is posed by the extensive use of water and of electrical appliances in the same work areas. Each member of the staff must be trained and alerted to the potential dangers, and electrical installations must be made by professionals who are aware of the nature of the facility and of this hazard.

Other hazards result from the prevalence of water and the height of cage racks. All floor and ladder surfaces should be constructed of or covered with "nonskid" materials and attention should be given to reducing the numbers of sharp corners or devices that might prove dangerous should personnel fall against them. The use of "nonskid" footwear is also recommended.

2. Biological

Diseases transmitted from animals to man are known as zoonoses. The peril to humans by contact with North American amphibians is considered so minimal that experiments with living anurans and urodeles constitute part of the biology curriculum of almost all high schools, colleges, and universities. However, researchers and teachers should be aware that most amphibians obtained from commercial sources are wild-caught (see Chapter III, Section B.2) and, like all animals taken from nature and many household pets, may harbor etiological agents for diseases infectious to man (Van der Hoeden, 1964; Schwabe, 1969; Benenson, 1970). Accordingly, it seems prudent to instruct all personnel working with amphibians to wear protective gloves if they have cuts and cracks on their hands, not to touch mucous membranes while working, and to wash their hands carefully after handling amphibians.

Chapter IX, Section C describes the herpesvirus associated with a prevalent renal tumor that afflicts *R. pipiens*. There is no evidence at this time to suggest that humans are susceptible to infection by frog viruses that would result in a neoplasm or any other disease. While this is true, laboratory staff exposed to frog viruses should be alerted to possible risks and methods of avoiding them. It is recommended that personnel who are exposed to tumor virus preparations be trained in microbiological technique and in the correct handling and disposal of infected animals. Particular precautions should be observed by pregnant staff (Chesterman, 1967). Hazards from working on oncogenic viruses and preventive measures that afford protection are discussed in a manual edited by Hellman (1969) and in a recent book by Hellman, Oxman, and Pollack (1973). The former should be read by all personnel prior to beginning work with any oncogenic virus.

Workers should also be alerted that some exotic amphibians produce highly dangerous skin secretions (Albuquerque *et al.*, 1971; Daly and Witkop, 1971), and some produce secretions that, although not toxic, may be irritating. Personnel should be warned when such animals are brought to the laboratory, the animals should be clearly labeled, and the use of protective gloves and careful washing specifically encouraged.

The use of live insects or insect products as well as the presence of animals that splash lead to the production of dusts and aerosols that may be allergenic. Personnel should be made aware of this, particularly in reference to diagnosing skin and nasopharyngeal conditions. Although few instances have been reported, allergic responses in amphibian quarters may be expected as the number of installations increases.

Appendix A Status of "Endangered" Amphibians

The information presented has been assembled from the *Red Data Book: Volume 3,* Amphibia and Reptilia (1968 and 1970 data sheets) compiled by René E. Honegger and published by the Survival Service Commission of the International Union for Conservation of Nature and Natural Resources, and the Endangered Species List (Department of the Interior).

Species	Source	Status Category[a]
Megalobatrachus japonicus japonicus (Japanese giant salamander)	Japan	2(b)
Megalobatrachus japonicus davidianus (Chinese giant salamander)	China	4(b)
Ambystoma dumerili dumerili (Lake Patzcuaro salamander-Achoque)	Mexico	2(b)R
Ambystoma lermaensis (Lake Lerma salamander)	Mexico	2(a)R
Ambystoma macrodactylum croceum (Santa Cruz long-toed salamander)	California	1(b)
Ambystoma tigrinum californiense (California tiger salamander)	California	3(b)
Chiglossa lusitanica (goldstriped salamander)	Spain and Portugal	4(a)
Plethodon larselli (Larch Mountain salamander)	Oregon	4(a)R
Typhlomolge rathbuni (Texas blind salamander)	Texas	1(a)RM
Hydromantes shastae (shasta salamander)	California	2(a)R
Hydromantes brunus (Limestone salamander)	California	4(a)R
Leiopelma archeyi (Archey's frog, Coromandel Leiopelma)	New Zealand	2(a)PR

Leiopelma hamiltoni	New Zealand	2(a)PR
(Hamilton's frog, Stephens Island Leiopelma)		
Leiopelma hochstetteri	New Zealand	2(a)PR
(Hochstetter's frog, North Island Leiopelma)		
Xenopus gilli	South Africa	4(a)R
(Cape Platana)		
Discoglossus nigriventer	Israel	1(a)P
(Israel painted frog or Israel discoglossus)		
Bufo boreas nelsoni	United States	4(b)
(Amargosa toad)		
Bufo boreas exsul	California	2(b)
(black toad, Inyo County toad)		
Bufo houstonensis	Texas	1(a)(***)
(Houston toad)		
Bufo retiformis	United States	4(a)T
(Sonoran green toad)		
Hyla andersoni	United States	1(a)R
(Pine Barrens tree frog)		
Pseudacris streckeri illinoensis	United States	4(b)
(Illinois chorus frog)		
Nesomantis thomasseti	Seychelle Is.	4(a)R
(Seychelle Islands frog)		
Sooglossus seychellensis	Seychelle Is.	4(a)R
(Seychelle Islands frog)		
Sooglossus gardinieri	Seychelle Is.	4(a)R
(Seychelle Islands frog)		
Rana pipiens fisheri	Nevada	1(b)
(Vegas Valley leopard frog)		
Megalixalus seychellensis	Seychelle Is.	4(a)R
(Seychelle Islands tree frog)		

[a]Classification of rare and endangered forms:

Category 1 ENDANGERED In immediate danger of extinction: continued survival unlikely without the implementation of special protective measures.

Category 2 RARE Not under immediate threat of extinction, but occurring in such small numbers and/or in such a restricted or specialized habitat that it could quickly disappear.

Category 3 DEPLETED Although still occurring in numbers adequate for survival, the species has been heavily depleted and continues to decline at a rate that gives cause for serious concern.

Category 4 INDETERMINANT Apparently in danger, but insufficient data currently available on which to base a reliable assessment of status. Needs further study.

Star listing ***Species or subspecies critically endangered.

Symbols (a) Full species.
    (b) Subspecies.
    E Exotic, introduced or captive populations believed more numerous than indigenous stock.
    M Under active management in a national park or other reserve.
    P Legally protected, at least in some parts of its range.
    R Included because of its restricted range.
    S Secrecy still desirable.
    T Subject to substantial export trade.

Appendix B Control Laws by States and Canadian Provinces

These data were collected by a survey conducted by Nace in 1970 and from the statutes. Assembly was accomplished through the assistance of Stafford Cox, student of Environmental Law at the University of Michigan.

UNITED STATES

1. **ALABAMA**   **No regulations**
    Department of Conservation, Game and Fish Division, Montgomery, Alabama 36104

2. **ALASKA**   **No regulations**
    Department of Fish and Game, Subport Building, Juneau, Alaska 99801
    Comment   The major amphibian is *Rana sylvatica*. It is doubtful there is any commercial harvesting of this amphibian.
    Reference   Division of Biological Sciences, University of Alaska, College, Alaska 99701

3. **ARIZONA**   **Regulations**
    Game and Fish Department, Arizona State Building, Phoenix, Arizona 85007
    Species   *Rana catesbeiana*
    Season   June 1–November 30
    Commercial   A valid fishing license is required.
    Research   A scientific collecting permit is available.
    Take   12/day
    Possession   12
    Techniques   Day or night; using spear, gig, hook and line, bow and arrow, or artificial light; but no explosive, firearm, net, or trap
    Comment   No other amphibians are protected by law.
    Reference   Game and Fish Commission Order, Number T-41, Number T-42, *Date* December 13, 1969

4. **ARKANSAS**   **Regulations**
    Game and Fish Commission, Little Rock, Arkansas 72201
    Species   Frogs
    Season   April 16–December 31
    Commercial   A valid fishing license is required.

Take 18/day
Possession 36
Techniques Frogs may be taken only by the use of hand, handnet, hook and line, gig, spear, or bow and arrow.
Breeders Frogs legally acquired may be raised in captivity for propagation and sale under a permit that costs $25.00/year. Some species may be raised for scientific and restocking purposes only under a permit that costs $2.50/year.
Comment It is unlawful to ship, take, transport, or export out of the state for sale any *Rana catesbeiana* or parts thereof in any manner whatsoever, except those raised by a game breeder or fish farmer.
Reference 4A-ASA, Title 47, § 508, 509 (1947)

5. CALIFORNIA — Regulations
Department of Fish and Game, 1416 9th Street, Sacramento, California 95814
Species All edible frogs regardless of size
Season Southern California, June 1–November 30; central and northern California, all year
Commercial A valid fishing license is required.
Research A permit of $5.00/year is required.
Possession Southern California, 12; central and northern California, 24
Techniques Day or night; using light, spear, gig, grab, paddle, hook and line, dip net, bow and arrow, fishing tackle, or hand
Reference 31-CAC, Div. 6, Ch. 7, § 6850-55, 6880-85 (1958)

6. COLORADO — Regulations
Game, Fish, and Parks Division, 6060 Broadway, Denver, Colorado 80216
Species *Rana catesbeiana*
Season August 1–September 30
Commercial No more than 5,000 participants/season. One-quarter ton of urodeles exported/year by bait dealers.
Take 10/day
Comment Several amphibian suppliers have been discouraged due to fact that 89.7% of waters are contained by man-made impoundments.
No other amphibians are protected by law.
Reference 4-CRS, Ch. 62, § 1.1, 3.4 (1963)

7. CONNECTICUT — No regulations
Board of Fisheries and Game, State Office Building, Hartford, Connecticut 06101
Commercial Bait species are protected by law and a fishing or hunting license is required.
Comment The commission may remove undesirable plants or animals from the waters by chemical, electrical, or mechanical means.
Reference 13-CGSA, Title 26, § 22, 27 (1973)

8. DELAWARE — Regulations
Board of Game and Fish Commissioners, P.O. Box 457, Dover, Delaware 19901
Species *Rana catesbeiana*
Season May 1–December 31
Commercial A valid hunting or fishing license is required.
Take With a fishing license, 10/day; with a hunting license, 24/day
Reference 3-DCA, Title 7, § 701, 703, 704, 710 (1953)

9. FLORIDA — Regulations
Game and Fresh Water Fish Commission, Farris Bryant Building, 620 South Meridan Street, Tallahassee, Florida 32304

Species Frogs
Season All year (except on wildlife management areas)
Research A permit is available for wildlife management areas.
Techniques May be taken by gig, club, blow gun, hand, hook and line, or by shooting during daylight hours only.
Comment Florida's laws relating to amphibians are designed primarily for the protection of animals other than amphibians.
Reference 1970 *Wildlife Code of the State of Florida;* 12-FSA, Title 18, Ch. 285.15 (1962)

10. GEORGIA   **No regulations**
Game and Fish Commission, 401 State Capitol, Atlanta, Georgia 30334

11. HAWAII   **Regulations**
Division of Fish and Game, 400 S. Beretania Street, Honolulu, Hawaii 96813
Species Frogs greater than 9 in. from snout to extended hind feet
Season All year
Comment Small frogs less than 9 in. from snout to extended hind feet and tadpoles are protected. It is unlawful for any person to catch, take, kill, expose, to have in possession, or to offer for sale any tadpole or small frog.
Violation Any person who violates these regulations shall be fined not less than $5.00 and not more than $50.00.
Reference 3-HRS, Title 12, Ch. 188, § 62–67 (1968)

12. IDAHO   **Regulations**
Fish and Game Department, P.O. Box 25, Boise, Idaho 83707
Species Rana catesbeiana
Season All year
Commercial A valid fishing license is required. A permit is necessary to sell bull frogs.
Take 12/day
Possession 12
Techniques May use bow and arrow, spear, or mechanical devices.
Comment No other amphibians are protected by law.
Reference 1970 *Idaho Fishing Seasons and Regulations*

13. ILLINOIS   **Regulations**
Department of Conservation, State Office Building, Springfield, Illinois 62706
Species Rana catesbeiana
Season June 15–August 31
Commercial A valid fishing or hunting license is required.
Take 8/day
Possession 8
Comment Article II, Fish Protection Regulations, Ownership and Title, § 20 "If any person causes any waste or sewage to be discharged into or causes or allows pollution of any waters of this State so as to kill aquatic life, the Department, through the Attorney General, may bring an action against such person and recover the reasonable value of the aquatic life destroyed by such waste, sewage, or pollution. Any money so recovered shall be placed in the Game and Fish Fund in the State Treasury."
Reference *Fish Code of Illinois,* Article 2, § 15, 16, 20, 21, 23 (1969 edition); 19-ILP, Title Fish and Game, p. 457 (1968)

14. INDIANA   **Regulations**
Division of Fish and Game, Room 605 State Office Building, Indianapolis, Indiana 46209

Species Frogs
Season June 15-April 30
Commercial A valid hunting and trapping license is required.
Take 25/day
Possession After 2 days, no more than 50 frogs
Techniques May use gig, spear (head not more than 3 in. in width and a single row of tines), long bow and arrow, air rifle, .22 caliber firearm, club, or hands. With not more than one pole or hand line, and these shall not have more than one hook or artificial lure.
Comment No other amphibians are protected by law.
Reference DNS Indiana Discretionary Order No. W-16; 4-ISA, Title 11, § 1643 (1956)

15. IOWA — Regulations
Conservation Commission, 300 Fourth Street, Des Moines, Iowa 50319
Species Frogs of the family Ranidae
Season May 12-November 30
Commercial A valid fishing license is required.
Take 48/day
Possession 96 (A licensed bait dealer may have 240.)
Comment No person shall prevent frogs from having free access to and egress from water. Frog export is prohibited. There are no regulations enforced upon private landowners.
Reference 7-ICA, Title 5, Ch. 109, § 12, 13, 32, 38, 45, 63, 67, 84, 85, 119 (1946)

16. KANSAS — Regulations
Forestry, Fish, and Game Commission, Box 1028, Pratt, Kansas 67124
Species Rana catesbeiana
Season July 1-September 30
Commercial A valid fishing license is required.
Take 8/day
Techniques May use dip net, hook and line, or hand.
Reference 3-KSA, Ch. 32, § 147, 148, 149, 161, 162, 164 (1964)

17. KENTUCKY — Regulations
Fish and Wildlife Resources Department, State Office Building, Annex, Frankfort, Kentucky 40601
Species Salamanders, tadpoles, and frogs
Commercial A valid live fish and bait license is required.
Possession 100 salamanders, 100 tadpoles, frogs unlimited
Techniques May hunt at night with lights to blind frogs.
Reference 7-KRS, Ch. 150, § 175, 360 (1971)

18. LOUISIANA — Regulations
Wildlife and Fisheries Commission, 400 Royal Street, New Orleans, Louisiana 70130
Species Bullfrogs and lagoon frogs
Season June 1-March 31
Commercial A permit is necessary during closed season.
Research A person may take any species of small frogs, green, grass, leopard, or spring frogs for scientific, educational, or propagating purposes, regardless of size, but not for food or sale.
Reference 28-LRS, Title 56, Ch. 1, § 328, 363, 369, 375, 384 (1964)

19. **MAINE**   **No regulations**
Department of Inland Fisheries and Game, State House, Augusta, Maine 04330

20. **MARYLAND**   **No regulations**
Game and Inland Fish Department, State Office Building, Annapolis, Maryland 21404

21. **MASSACHUSETTS**   **No regulations**
Division of Fisheries and Game, 100 Cambridge Street, Boston, Massachusetts 02202

22. **MICHIGAN**   **Regulations**
Department of Conservation, Stevens T. Mason Building, Lansing, Michigan 48926
Species Frogs
Season May 30–November 15
Commercial No license is required to take or sell.
Research A scientific collector's permit is available for November 16–May 29.
Techniques May not spear with artificial light.
Reference 9-MSA, Title 13, § 1711, 1712, 1713, 1716 (1967)

23. **MINNESOTA**   **Regulations**
Department of Conservation, Centennial Office Building, St. Paul, Minnesota 55101
Species Frogs
Season May 16–March 31
Commercial A valid fishing license is required.
Research A permit may be issued for April 1–May 15.
Possession Unlimited when less than 6 in.; 150 frogs when greater than 6 in. in length from nose to tip of hind feet toes.
Comment Commissioner shall establish regulations dealing with purchase, possession, and transportation of frogs for purposes other than bait.
    In the year of 1969: 416 persons were licensed to take, possess, transport, sell frogs for purposes other than bait; 9 persons, resident, received frog dealer license; and 5 persons, nonresident, received frog dealer license.
Reference 8-MSA, Ch. 101.44 (1968)

24. **MISSISSIPPI**   **Regulations**
Game and Fish Commission, Box 451, Jackson, Mississippi 39205
Species Frogs
Season All year
Commercial A hunting and fishing license is required.
Take 25/day
Possession 50
Reference Digest of Mississippi Hunting and Trapping Regulations (1971)

25. **MISSOURI**   **Regulations**
Missouri Conservation Commission, P.O. Box 180, Jefferson City, Missouri 65101
Species Rana catesbeiana
Season June 30–November 30
Commercial A valid fishing or hunting license is required.
Research A scientific collector's permit is available. Frogs may also be propagated and sold with a fish farming or fish hatchery permit.
Take 8/day
Possession 8

Techniques With a fishing permit, one may use hand, gig, handnet, long bow, hook and line, or artificial light. With a hunting permit, one may use .22 caliber rimfire rifle, pistol, or pellet gun.

Comment It is illegal to sell other amphibians of any kind under any conditions. It is suggested that the best approach to the biological supply problem would be artificial propagation.

Reference 2-MRS, Ch. 252.020 (1969)

26. MONTANA — No regulations
Fish and Game Department, Helena, Montana 59601

27. NEBRASKA — Regulations
Game Commission, Lincoln, Nebraska 68509

Species Rana catesbeiana
Season July 1–November 1
Commercial A bait vendor permit is required.
Take 8/bag
Techniques With a hunting permit, one may use firearms, bow and arrow, hand, or handnet. With a fishing permit, one may use hand, handnet, gig, hook and line, or artificial light.
Reference 3-NRS, Ch. 37, § 226, 503, 505 (1943)

28. NEVADA — Regulations
Fish and Game Commission, Box 10678, Reno, Nevada 89510

Species Rana catesbeiana
Commercial A valid fishing license is required.
Take Douglas and Esmeralda counties, 10/day; all other counties, 15/day
Possession 15
Endangered Vegas Valley leopard (*Rana pipiens fisheri*)
Techniques May use hand, bow and arrow, hook and line, gig, or spear. Firearms are not allowed.
Reference 16-NRS, Title 45, Ch. 501.037 and 503.290 (1971)

29. NEW HAMPSHIRE — Regulations
Fish and Game Department, Concord, New Hampshire 03301

Comment A permit is required.

30. NEW JERSEY — No regulations
Division of Fish and Game, Box 1809, Trenton, New Jersey 08625

31. NEW MEXICO — Regulations
Game and Fish Department, State Capitol, Santa Fe, New Mexico 87501

Species Rana catesbeiana
Season August 1–August 31
Commercial A valid fishing license is required.
Research A scientific collector's permit is available.
Take 8/day
Possession 8
Reference NMSA, Title 53, Ch. 2, § 4, 18, 23 (1970)

32. NEW YORK — Regulations
New York State Department of Environmental Conservation, Albany, New York 12201

Species Frogs
Season June 16–September 30
Commercial A valid fishing or hunting license is required.
Techniques With a fishing license, one may use hand, spear, club, hook and line. With a hunting license, one may use gun or long bow.

Comment No person shall use any device that prevents frogs from having free access to and egress from the water.

Reference 10-CLNYA, Article 4, § 181, 210, 225, 228, 332, 387 (1967)

33. NORTH CAROLINA — Regulations
Wildlife Resources Commission, Box 2919, Raleigh, North Carolina 27602

Comment A collector's permit is required.

Reference 3-GSNC, Ch. 113.91 (1963)

34. NORTH DAKOTA — Regulations
Game and Fish Department, Bismarck, North Dakota 58501

Species Frogs

Season June 1–March 31 (May hunt from April 1–May 31 if for angling purposes, academic study, or with a commercial frog license.)

Possession 12 dozen for bait. There is no limit for persons with commercial license.

Violation Guilty of misdemeanor

Comment The state has extensive populations of salamanders and scattered mudpuppies.

Reference 4-NDCCA, Ch. 20.12 (1970)

35. OHIO — Regulations
Natural Resources Department, Wildlife Division, Columbus, Ohio 43212

Species Frogs

Season June 15–May 1

Commercial A valid fishing license is required.

Take 10/day

Possession 10

Techniques May use bow and arrow, but not shooting guns.

Comment Field observation indicates that there is no overharvesting.

Reference ORCA, Title 15, Ch. 1533.32 (1964)

36. OKLAHOMA — Regulations
Wildlife Conservation Department, 1801 No. Lincoln, Oklahoma City, Oklahoma 73105

Commercial Production of fish and frogs

Reference OSA, Title 29, Ch. 1765 (1968)

37. OREGON — Regulations
State Game Commission, 1634 S.W. Alder Street, Portland, Oregon 97208

Species Rana catesbeiana

Season All year

Commercial A valid fishing license is required.

Possession 12

Techniques Day or night; one can not use firearms, air- or gas-operated guns, or explosives.

Comment There is one live fish bait supplier with bullfrogs.

Reference 4-ORS, Ch. 496.185, 496.190, 497.010, 497.020, 497.040, 497.050, 498.030, 498.095 (1953)

38. PENNSYLVANIA — Regulations
Pennsylvania Game Commission, P.O. Box 1567, Harrisburg, Pennsylvania 17120

Species Frogs and tadpoles

Season July 1–November 1

Commercial A valid fishing license is required.

Research A scientific collector's permit is available.

Take 15/day (frogs or tadpoles)

Possession 15 (frogs or tadpoles)
Violation $10.00 for every frog or tadpole over 15
Techniques No lights at night ($25.00 fine)
Reference PSA, Title 30, Ch. 2, § 321-329 (1958)

39. RHODE ISLAND — No regulations
Department of Natural Resources, Veterans Memorial Building, Providence, Rhode Island 02903

40. SOUTH CAROLINA — No regulations
Division of Game, Box 360, Columbia, South Carolina 29202

41. SOUTH DAKOTA — Regulations
Department of Game, Fish, and Parks, Pierre, South Dakota 57501
Species Frogs
Season May 2–October 15
Commercial Frog farm and dealer licenses are available.
　1967 19,544 lb wholesale; 5,242 dozen for bait
　1968 5,175 lb wholesale; 507 dozen for bait
Techniques Cannot use firearms.
Comment Cannot export frogs less than 3 in. long.
Reference 12-SDCLA, Title 41, Ch. 6, § 46-48, and Ch. 14, § 31 (1967)

42. TENNESSEE — Regulations
Game and Fish Commission, 706 Church Street, Nashville, Tennessee 37203
Species Rana catesbeiana
Season All year
Commercial A collector's or dealer's permit is available. One can sell only raised or imported bullfrogs.
Research A scientific collecting permit is available.
Take 10/day
Possession 20
Techniques No firearms may be used in wildlife management and refuge areas defined as hunting.
Reference Tennessee Game and Fish Commission, Proclamation No. 215.

43. TEXAS — No regulations
Parks and Wildlife Department, Austin, Texas 78701
Comment Rana pipiens, Rana catesbeiana, and *Necturus* may be taken in any numbers and at any time.

44. UTAH — No regulations
Fish and Game Division, 1596 West North Temple, Salt Lake City, Utah 84116

45. VERMONT — No regulations
Fish and Game Department, Montpellier, Vermont 05602
Comment There is one commercial dealer.

46. VIRGINIA — Regulations
Commission of Game and Inland Fisheries, Box 1642, Richmond, Virginia 23213
Comment A valid hunting license is required.

47. WASHINGTON — Regulations
Department of Game, 600 N. Capitol Way, Olympia, Washington 98502
Species Rana catesbeiana
Season At times and in waters that are open for the taking of fish.
Commercial A valid hunting or fishing license is required.
Take 10/day

Possession 10
Comment Waterdogs are prohibited in Silverlake and Cowlitz counties.
Referral *1971 Game Fish Seasons and Catch Limits*

48. WEST VIRGINIA — Regulations
Department of Natural Resources, Charleston, West Virginia 25305
Species Frogs
Season June 13–July 31
Commercial A license is available to raise and sell frogs.
Research A permit is issued with a written application explaining details, methods, and purpose. No charge.
Take 10/day
Possession 20 with no size limitations, cannot have more than 100 total aquatic animals.
    A survey of licensed fishermen on gigging showed that in 1968: 113,222 frogs were gigged.
Comment In state: salamanders, 25 species; toads, 3 species; frogs, 11 species. These species are the property of the state, except on private land.
Reference 8-WVC, 20.2.11, 20.2.48 (1966)

49. WISCONSIN — Regulations
Conservation Department, Box 450, Madison, Wisconsin 53701
Species Frogs
Season May 2–December 31
Commercial May sell out of season if caught in season or raised.
Techniques No firearms may be used.
Reference *Wisconsin Fishing Summary of 1970 Regulations*

50. WYOMING — No regulations
Game and Fish Commission, Box 1589, Cheyenne, Wyoming 82001
Comment Frogs are included in wildlife, which is the property of the state.

CANADIAN PROVINCES

1. ALBERTA — No regulations
Fish and Wildlife Division, Department of Lands and Forests, Edmonton, Alberta
Reference 19-*Statutes of Alberta,* Ch. 113 (1970)

2. BRITISH COLUMBIA — No regulations
Fish and Wildlife Branch, Department of Recreation and Conservation, Victoria, British Columbia
Reference 2-*Revised Statutes of British Columbia,* "The Game Act," Ch. 160 (1960)

3. MANITOBA — Regulations
Department of Mines, Resources, and Environmental Management, Box 11, Building 15, 139 Tuxedo Avenue, Winnipeg, Manitoba, R3C OV8
Species Tiger salamander (*Ambystoma tigrinum*), leopard frog (*Rana pipiens*), Plains spadefoot toad (*Scaphiophus bombifrons*), green frog (*Rana clamitans*), mink frog (*Rana septentrionalis*)
Season Leopard frogs may be caught and sold from June 1 to September 30, 1973, or until 100,000 lb have been sold, whichever comes first.
    Tiger salamanders may be caught from June 1 to September 30, 1973, with no limit on numbers taken.

Plains spadefoot toad, green frog, and mink frog may be taken for scientific or educational purposes only under authority of a scientific collecting permit. They may be taken by others, e.g., fanciers, without a permit but may not be sold.

Commercial Special licenses are required for "pickers" and "dealers."

Take A report of amounts bought and sold is required from each dealer on the 15th and 30th of each month.

Comment "The Wildlife Act" amended in 1970 to include amphibians in the definition of wildlife.

Reference Manitoba Regulation 134/71 authorized under "The Wildlife Act," *Statutes of Manitoba,* Ch. W140; also Manitoba Regulations 81/73, 82/72, and 76/73.

4. NEW BRUNSWICK — No regulations

Fish and Wildlife Branch, Natural Resources Department, Fredericton, New Brunswick

Reference 2-*Revised Statutes of New Brunswick,* "The Game Act," Ch. 95 (1952)

5. NEWFOUNDLAND — No regulations

Federal Department of Fisheries, Box 5667, St. Johns, Newfoundland
Director of Wildlife, Department of Mines, Agriculture and Resources, St. Johns, Newfoundland

Comment *Rana clamitans* are present on the Island of Newfoundland. Also, frogs and toads exist in Labrador.

Reference 4-*Revised Statutes of Newfoundland,* Ch. 197, Title XX, "The Wildlife Act" (1952)

6. NOVA SCOTIA — No regulations

Conservation Division, Department of Lands, and Forests, Box 516, Kentville, Nova Scotia

7. ONTARIO — Regulations

Department of Lands and Forests, Parliament Buildings, Toronto, Ontario

Species *Rana catesbeiana*

Season July 1–October 15

Commercial A valid license is required at a fee of $5.00.

Take No person other than the holder of a license shall take more than 25 bullfrogs/day.

Possession No person other than the holder of a license shall possess more than 25 bullfrogs at one time.

Comment The regulations are in force only in the counties of Lanark and Leeds.

Reference Regulation 359 under "The Game and Fish Act," 2-*Revised Statutes of Ontario,* 186:1, 73, 74, 75, 91, 92 (1970)

8. PRINCE EDWARD ISLAND — No regulations

Fish and Wildlife Division, Department of Fisheries, Charlottetown, Prince Edward Island

9. QUEBEC — No regulations

Department of Tourism, Fish and Game, Quebec City, Quebec

Reference *Statutes of Quebec,* "Wildlife Conservation Act," Ch. 58 (1969)

10. SASKATCHEWAN — No regulations

Department of Natural Resources, Regina, Saskatchewan

Reference *Statutes of Saskatchewan,* "The Animal Protection Act," Ch. 5 (1972)

References

Abrams, G. D. 1969. Diseases in an amphibian colony, p. 419–428. *In* M. Mizell [ed.] Biology of amphibian tumors. Springer-Verlag, New York.
Albuquerque, E. X., J. W. Daly, and B. Witkop. 1971. Batrachotoxin: Chemistry and pharmacology. Science 172:995–1002.
American Public Health Association. 1965. Standard methods for the examination of water and waste water, 12th ed. American Public Health Association, New York. 626 p.
Anonymous. 1973. Where have all the frogs gone? Mod. Med. 41(23):20–24.
Arnold, S. J. 1972. Evolution of courtship behavior in salamanders. 2 vol. Ph.D. thesis, University of Michigan, Ann Arbor.
Asher, Jr., J. H. 1970. Parthenogenesis and genetic variability. II. One-locus models for various diploid populations. Genetics 66:369–391.
Asher, Jr., J. H. in press a. Systems of reproduction. I. Optimal methods for producing isogenetic and congenic strains. Theor. Appl. Genet.
Asher, Jr., J. H. in press b. Systems of reproduction. II. The influence of linkage and fitness upon the genetic structure of automictic parthenogenesis populations. Theor. Pop. Biol.
Asher, Jr., J. H., and G. W. Nace. 1971. The genetic structure and evolutionary fate of parthenogenetic amphibian populations as determined by Markovian analysis. Am. Zool. 11:381–398.
Atz, J. W. 1971. Aquarium fishes. Viking, New York. 110 p.
Bachmann, B. T., E. A. Adelberg, and R. Bakerman. 1973. SAM: The "search and match" computer program of the *Escherichia coli* Genetic Stock Center. BioScience 23:35–36.
Bagnara, J. T., and H. L. Stackhouse. 1973. Observations on Mexican *Rana pipiens*. Am. Zool. 13:139–143.
Ballard, S., M. A. Griffity, and G. Controni. 1964. Part 2. The morphology and biochemical reactions of the *Moraxella-Mima* group. Am. J. Med. Tech. 30:263–269.
Balls, M., and L. N. Ruben. 1967. The transmission of lymphosarcoma in *Xenopus laevis*, the South African clawed toad. Cancer Res. 27:654–659.
Balls, M., and L. N. Ruben. 1968. Lymphoid tumors in amphibia: A review. Prog. Exp. Tumor Res. 10:238–260.
Bardach, J. E., J. H. Ryther, and W. O. McLarney. 1972. Aquaculture. Wiley-Interscience, New York. 868 p.

Beetschen, J-C. 1971. Thermosensibilité de la mutation "ascite caudale" chez *Pleurodeles waltlii* (amphibien urodèle). C.R. Acad. Sci. Paris 273:97–100.
Benenson, A. S. [ed.]. 1970. Control of communicable diseases in man, 11th ed. American Public Health Association, New York. 316 p.
Bennett, G. W. 1962. Management of artificial lakes and ponds. Reinhold, New York. 283 p.
Berns, M. W. 1965. Mortality caused by kidney stones in spinach-fed frogs (*Rana pipiens*). BioScience 15:297–298.
Bishop, S. C. 1947. Handbook of salamanders. Comstock, Ithaca, New York. 555 p.
Blackler, A. W., and M. Fischberg. 1968. Hybridization of *Xenopus laevis petersi* (poweri) and *X. l. laevis*. Rev. Suisse Zool. 75:1023–1032.
Blair, W. F. [ed.]. 1972. Evolution of the genus *Bufo*. University of Texas Press, Austin. 459 p.
Boterenbrood, E. C. 1966. Newts and salamanders, Ch. 52, p. 867–892. *In* UFAW handbook on the care and management of laboratory animals, 32nd ed. Livingston, London.
Boyer, Jr., C. I., K. Blackler, and L. E. Delanney. 1971. *Aeromonas hydrophilia* infection in the Mexican axolotl, *Siredon Mexicanum*. Lab. Anim. Sci. 21:372–375.
Briggs, R. 1952. An analysis of the inactivation of the frog sperm nucleus by toluidine blue. J. Gen. Physiol. 35:761–780.
Briggs, R. In press. Developmental genetics of the axolotl. F. Ruddle [ed.] 31st symposium of the Society for Developmental Biology. Academic Press, New York.
Briggs, R. and T. J. King. 1952. Transplantation of living nuclei from blastula cells into enucleated frogs' eggs. Proc. Natl. Acad. Sci., U.S. 38:455–463.
Brown, A. L. 1970. The African clawed toad. Butterworth, London. 120 p.
Brown, L. E. 1973. Speciation in *Rana pipiens* complex. Am. Zool. 13:73–79.
Cairns, A. M., J. W. Bock, and F. G. Bock. 1967. Leopard frogs raised in partially controlled conditions. Nature 213:191–193.
Chesterman, F. C. 1967. Viruses, p. 60–67. *In* R. W. Raven and F. C. Roe [ed.] The prevention of cancer. Appleton-Century-Crofts, New York.
Cicmanec, J. L., D. L. Ringler, and E. S. Benke. 1973. Spontaneous occurrence and experimental transmission of fungus, *Fonsecaea pedrosoi*, in the marine toad, *Bufo marinus*. Lab. Anim. Sci. 23:43–47.
Clark, J. R., and R. L. Clark [ed.]. 1971. Sea water systems for experimental aquariums. T. F. H. Publications, Neptune City, New Jersey. 192 p.
Clark, H. F., C. Gray, F. Fabian, R. F. Zeigel, and D. T. Karzon. 1969. Comparative studies of amphibian cytoplasmic virus strains isolated from the leopard frog, bullfrog, and newt, p. 310–326. *In* M. Mizell [ed.] Biology of amphibian tumors. Springer-Verlag, New York.
Collard, W., H. Thorton, M. Mizell, and M. Green. 1973. Virus-free adenocarcinoma of the frog (summer phase tumor) transcribes Lucké tumor herpesvirus-specific RNA. Science 181:448–449.
Committee on Animal Nutrition, National Research Council. 1972. Number 10: Nutrient requirements of laboratory animals, 2nd ed. National Academy of Sciences, Washington, D.C. 117 p.
Committee on Revision of the Guide for Laboratory Animal Facilities and Care. 1972. Guide for the care and use of laboratory animals. U.S. Dep. Health, Educ., and Welfare, National Institutes of Health, Bethesda, Md. 56 p.
Committee on Standards, Institute of Laboratory Animal Resources. In press. Fishes: Guidelines for the breeding, care, and management of laboratory animals. National Academy of Sciences, Washington, D.C.

Conant, R. 1958. A field guide to reptiles and amphibians. Houghton-Mifflin, Boston. 366 p.
Cooke, A. S. 1971. Selective predation by newts on frog tadpoles treated with DDT. Nature 229:275-276.
Cooke, A. S. 1972. Indications of recent changes in status in the British Isles of the frog (*Rana temporaria*) and the toad (*Bufo bufo*). J. Zool. Proc. Zool. Soc. London 167:161-178.
Crans, W. J. 1969. Preliminary observations on frog filariasis in New Jersey. Proc. Annu. Conf. Bull. Wildl. Dis. Assoc. 5:342-347.
Culley, Jr., D. D. 1973. Use of bullfrogs in biological research. Am. Zool. 13:85-90.
Culley, Jr., D. D., and S. P. Meyers. 1972. Frog culture and ration development. Feedstuffs 44(31):26.
Cullum, L., and J. T. Justus. 1973. Housing for aquatic animals. Lab. Anim. Sci. 23: 126-129.
Culp, R. L., and G. L. Culp. 1971. Advanced wastewater treatment. Van Nostrand Reinhold, New York. 310 p.
Daly, J. W., and B. Witkop. 1971. Chemistry and pharmacology of frog venoms, p. 497-519. *In* W. Bücherl, E. E. Buckley and V. Deulofeu [ed.] Venomous animals and their venoms, Vol. 2. Academic Press, New York.
Dasgupta, S. 1962. Induction of triploidy by hydrostatic pressure in the leopard frog, *Rana pipiens*. J. Exp. Zool. 151:105-121.
Davidson, E. H., and B. R. Hough. 1969. Synchronous oogenesis in *Engystomops pustulosus*, a neotropic anuran suitable for laboratory studies: Localization in the embryo of RNA synthesized at the lampbrush stage. J. Exp. Zool. 172: 25-48.
Dawe, C. J. 1969. Some comparative morphological aspects of renal neoplasms in *Rana pipiens* and of lymphosarcomas in amphibia, p. 429-440. *In* M. Mizell [ed.] Biology of amphibian tumors. Springer-Verlag, New York.
Deuchar, E. M. 1972. *Xenopus laevis* and developmental biology. Biol. Rev. 47: 37-112.
Di Berardino, M. A. 1967. Frogs, p. 53-74, *In* F. H. Wilt and N. K. Wessells [ed.] Methods in developmental biology. Thomas Y. Crowell Co., New York.
Duryee, W. R., M. E. Long, H. C. Taylor, Jr., W. P. McKelway, and R. L. Ehrmann. 1960. Human and amphibian neoplasms compared. Science 131:276-280.
D'Ver, A. S. 1973. The animal lab's lost science: Personnel management. Lab. Anim. 2:16.
Ecker, A., and Wiedersheim, R. 1896. Anatomie des Frosches. Friedrich Vieweg und Sohn, Braunschweig.
Elkan, E. 1960. Some interesting pathological cases in amphibians. Proc. Zool. Soc. London 134:375-396.
Elsdale, T. R., J. B. Gurdon, and M. Fischberg. 1960. A description of the technique for nuclear transplantation in *Xenopus laevis*. J. Embryol. Exp. Morphol. 8:437-444.
Emerson, H., and C. Norris. 1905. "Red-leg," an infectious disease of frogs. J. Exp. Med. 7:32-58.
Emery, A. R., A. H. Berst, and K. Kodaira. 1972. Under-ice observation of wintering sites of leopard frogs. Copeia 1:123-126.
Emmons, M. B. 1973. Problems of an amphibian supply house. Am. Zool. 13:91-96.
Ewing, W. H., R. Hugh, J. G. Johnson. 1961. Studies on the *Aeromonas*. U.S. Dep. Health, Educ., and Welfare, Communicable Disease Center, Atlanta, Georgia. 37 p.

Fankhauser, G. 1945. The effects of the change in chromosome number on amphibian development. Q. Rev. Biol. 20:20–78.

Fankhauser, G. 1955. The role of nucleus and cytoplasm, p. 126–150. In B. H. Willier, P. A. Weiss, and V. Hamburger [ed.] Analysis of development. W. B. Saunders Co., Philadelphia.

Farrar, E. S. 1972. Some aspects of carbohydrate metabolism and its regulation by adrenalin and glucagon in *Rana pipiens.* Ph.D. thesis, University of Michigan, Ann Arbor.

Farrell, R. K., and S. D. Johnson. 1973. Identification of laboratory animals: Freeze marking. Lab. Anim. Sci. 23:107–110.

Fraser, L. R. 1971. Physio-chemical properties of an agent that induces parthenogenesis in *Rana pipiens* eggs. J. Exp. Zool. 177:153–172.

Frazer, J. F. D. 1966. Frogs and toads, Ch. 51, p. 853–866. In UFAW handbook on the care and management of laboratory animals, 3rd ed. Livingston, London.

Freed, J. J., and L. Mezger-Freed. 1970. Stable haploid cultured cell lines from embryos. Proc. Natl. Acad. Sci., U.S. 65:337–344.

Gasche, P. 1943. Die Zucht von *Xenopus laevis* Daudin und ihre Bedeutung fur die biologische Forschung. Rev. Suisse Zool. 50:262–269.

Gasche, P. 1944. Beginn und Verlauf der Metamorphose bei *Xenopus laevis* Daudin Festlegung von Umwandlungsstadien. Helv. Physiol. Pharmacol. Acta 2:607–626.

Gay, W. I. (chairman). 1971. Defining the laboratory animal. IV Symposium, Int. Comm. on Laboratory Animals. National Academy of Sciences, Washington, D.C. 628 p.

Gibbs, E. L. 1963. An effective treatment for red-leg disease in *Rana pipiens.* Lab. Anim. Care 13:781–783.

Gibbs, E. L. 1973. *Rana pipiens* health and disease. How little we know. Am. Zool. 13:93–96.

Gibbs, E. L., T. J. Gibbs, and T. C. Van Dyck. 1966. *Rana pipiens*: Health and disease. Lab. Anim. Care 16:142–160.

Gibbs, E. L., G. W. Nace, and M. B. Emmons. 1971. The live frog is almost dead. BioScience 21:1027–1034.

Goin, C. J., and O. B. Goin. 1971. Introduction to herpetology, 2nd ed. Freeman, San Francisco. 356 p.

Gosner, K. L. 1960. A simplified table for staging anuran embryos and larvae with notes on identification. Herpetologia 16:183–190.

Granoff, A. 1969. Viruses of amphibia. Curr. Top. Microbiol. Immunol. 50:107–137.

Granoff, A. 1972. Lucké tumor-associated viruses–A review, p. 171–182. In P. M. Briggs, G. de-Thé, and L. N. Payne [ed.] Oncogenesis and herpesviruses. International Agency for Research on Cancer, Lyon.

Granoff, A., P. E. Came, and K. A. Rafferty, Jr. 1965. The isolation and properties of viruses from *Rana pipiens*: Their possible relationship to the renal adenocarcinoma of the leopard frog. Ann. N.Y. Acad. Sci. 126:237–255.

Gravell, M. 1971. Viruses and renal carcinoma of *Rana pipiens.* X. Comparison of herpes-type viruses associated with Lucké tumor-bearing frogs. Virology 43:730–733.

Gromko, M. H., S. M. Francene, and S. J. Smith-Gill. 1973. Analysis of the crowding effect in *Rana pipiens* tadpoles. J. Exp. Zool. 186:63–71.

Guillet, F., C-L. Gallien, and M-T Chalumeau-Le Foulgoc. 1971. Etude de la malate déshydrogenase du plasma sanguin et des hématies dans le genre *Pleurodeles* (amphibien, urodèle). C.R. Acad. Sci. Paris 272:1810–1812.

Gurdon, J. B. 1959. Tetraploid frogs. J. Exp. Zool. 141:519–543.
Gurdon, J. B. 1960. The effects of ultraviolet irradiation on uncleaved eggs of *Xenopus laevis*. Q. J. Microsc. Sci. 101:299–311.
Gurdon, J. B. 1967. African clawed frogs, p. 75–84. *In* F. H. Wilt and N. K. Wessells [ed.] Methods in developmental biology. Thomas Y. Crowell Co., New York.
Guyetant, R. 1964. Comparative action of light and dark on growth and metamorphosis of tadpoles of *Rana temporaria*. Zool. Physiol. 19:77–98.
Hadji-Azimi, I., and M. Fischberg. 1972. Some pathological aspects of the spontaneous lymphoid tumour in *Xenopus laevis*. Pathol. Microbiol. 38:118–132.
Hamburger, V. 1960. A manual of experimental embryology. University of Chicago Press, Chicago. 220 p.
Heatwole, H. 1961. Inhibition of digital regeneration in salamanders and its use in marking individuals for field studies. Ecology 42:593–594.
Hejmadi, P. M. 1970. Transfer of maternal serum proteins into the egg of *Rana pipiens* and their role in development. Ph.D. thesis. University of Michigan, Ann Arbor.
Hellman, A. [ed.]. 1969. Biohazard control and containment in oncogenic virus research. U.S. Dep. Health, Educ., and Welfare, National Institutes of Health, USPHS. Government Printing Office, Washington, D.C. 31 p.
Hellman, A., M. N. Oxman, and R. Pollack. 1973. Biohazards in biological research. Cold Spring Harbor Laboratory, New York. 230 p.
Hennen, S. 1970. Influence of spermine and reduced temperature on the ability of transplanted nuclei to promote normal development in eggs of *Rana pipiens*. Proc. Natl. Acad. Sci., U.S. 66:630–637.
Hirschfeld, W. J., C. M. Richards, and G. W. Nace. 1970. Growth of larval and juvenile *Rana pipiens* on four laboratory diets. Am. Zool. 10:315.
Holtfreter, J. 1931. Uber die Aufzucht isolierter Teile des Amphibien Keimes II. Arch. F. Ent. Mech. 124:404–465.
Hsu, C., and H. Liang. 1970. Sex races of *Rana catesbeiana* in Taiwan. Herpetologica 26:214–221.
Hutchison, V. H., and M. A. Kohl. 1971. The effect of photoperiod on daily rhythms of oxygen consumption in the tropical toad, *Bufo marinus*. Z. Vgl. Physiol. 75:367–382.
ILAR News. 1972. Annual survey of animals used for research purposes during calendar year 1971. 16(1):i–xv.
Inoue, S., M. Singer, and J. Hutchinson. 1965. Causative agent of a spontaneously originating visceral tumor of the newt, *Triturus*. Nature 205:408–409.
Jaeger, R. J. and R. J. Rubin. 1973. Extraction, localization, and metabolism of di-2-ethylhexyl phthalate from PVC plastic medical devices. Environ. Health Perspect., DHEW Publ. No. 73-218:95–102.
Jakowska, S. 1972. Fish protein concentrate as artificial diet for the tropical toad, *Bufo marinus*. Unpublished paper presented at the 23rd Annual AALAS Session, Oct. 1972.
Johnson, L. G., and E. P. Volpe. 1973. Patterns and experiments in developmental biology. Wm. C. Brown Company, Dubuque, Iowa. 255 p.
Joiner, G. N., and G. D. Abrams. 1967. Experimental tuberculosis in the leopard frog. J. Am. Vet. Assoc. 151:942–949.
Justus, J. T., and L. Cullum. 1971. A new method of housing axolotls and other aquatic amphibians. Lab. Anim. Sci. 21:110–111.

Kaess, W., and F. Kaess. 1960. Perception of apparent motion in the common toad. Science 132:953.

Kaplan, H. M. 1962. Toxicity of chlorine for frogs. Proc. Anim. Care Panel 12: 259-262.

Kaplan, H. M. 1969. Anesthesia in amphibians and reptiles. Fed. Proc. 28:1541-1546.

Kaplan, H. M., N. Yee, and S. S. Glaczenski. 1964. Toxicity of fluoride for frogs. Anim. Care 14:185-188.

Kaplan, H. M., and S. S. Glaczenski. 1965. Salamanders as laboratory animals: *Necturus.* Lab. Anim. Care 15(2):151-155.

Katagiri, C. 1961. On the fertilizability of the frog egg, I. J. Fac. Sci. Hokkaido Univ., Ser. VI, Zool. 14:607-613.

Kawamura, T., and M. Nishioka. 1963. Nucleo-cytoplasmic hybrid frogs between two species of Japanese brown frogs and their offspring. J. Sci. Hiroshima Univ., Ser. B., Div. 1 21:107-134.

Kawamura, T., and M. Nishioka. 1967. On the sex and reproductive capacity of tetraploids in amphibians. Gumma Symposia on Endocrinol. Proc. 4:23-39.

Kawamura, T., and M. Nishioka. 1972. Seven articles *in* Science report of the laboratory for amphibian biology, Hiroshima University. 1:1-350.

Kawamura, T., and M. Nishioka. 1973. Superiority of anuran amphibians as experimental materials. *In* Proceedings of the ICLA Asian Pacific Meeting on Laboratory Animals. Exp. Anim., Suppl. 22:11-126.

Keppler, W. J., W. Klassen, and J. B. Kitzmiller. 1965. Laboratory evaluation of certain larvicides against *Culex pipiens,* Linn, *Anopheles albimonus,* Wied, and *Anopheles quadrimaculatus,* Say. Mosquito News 25:415-419.

King, T. J. 1966. Nuclear transplantation in amphibia, p. 1-36. *In* D. M. Prescott [ed.] Methods in cell physiology, Vol. II. Academic Press, New York.

King, T. J. 1967. Amphibian nuclear transplantation, p. 737-751. *In* F. H. Wilt and N. K. Wessells [ed.] Methods in developmental biology. Thomas Y. Crowell Co., New York.

King, T. J., and R. Briggs. 1955. Changes in the nuclei of differentiating gastrula cells, as demonstrated by nuclear transplantation. Proc. Natl. Acad. Sci., U.S. 41:321-325.

Krauskopf, L. G. 1973. Studies on the toxicity of phthalates via ingestion. Environ. Health Perspect., DHEW Publ. No. 73-218:61-72.

Kulp, W. L., and D. G. Borden. 1942. Further studies on *Proteus hydrophilus,* the etiological agent in "red-leg" disease of frogs. J. Bacteriol. 44:673-685.

Littlejohn, M. L., and R. S. Oldham. 1968. *Rana pipiens* complex: Mating call structure and taxonomy. Science 162:1003-1004.

Loeb, J. 1899. On the nature of the process of fertilization and the artificial production of normal larvae (plutei) from the unfertilized eggs of the sea urchin. Am. J. Physiol. 3:135-138.

Lom, J. 1969. Cold-blooded vertebrate immunity to protozoa, p. 249-265. *In* G. J. Jackson, R. Herman, and I. Singer [ed.] Immunity to parasitic animals, Vol. 1. Appleton-Century-Crofts, New York.

Lunger, P. D. 1964. The isolation and morphology of the Lucké frog kidney tumor virus. Virology 24:138-145.

Lunger, P. D. 1966. Amphibia-related viruses. Adv. Virus Res. 12:1-33.

Mahoney, J. J., and V. H. Hutchison. 1969. Photoperiod acclimation and 24-hour variations in the critical thermal maxima of a tropical and a temperate frog. Oecologia 2:143-161.

Malacinski, G. H., and A. J. Brothers. In press. Mutant genes in the Mexican axolotl: A resource for the study of developmental genetics. Science.

Marlow, P. B., and S. Mizell. 1972. Incidence of Lucké renal adenocarcinoma in *Rana pipiens* as determined by histological examination. J. Natl. Cancer Inst. 48:823-829.

Masui, Y. 1967. Relative roles of the pituitary, follicle cells, and progesterone in the induction of oocyte maturation in *Rana pipiens*. J. Exp. Zool. 116:365-376.

Maugh, T. H. 1972. Frog shortage possible this winter. Science 178:387.

Mawdesley-Thomas, L. E. [ed.]. 1972. Diseases of fish. Academic Press, New York. 380 p.

McKee, J. E., and H. W. Wolf. 1963. Water quality criteria. Resources Agency, California State Water Resources Control Board, Sacramento. Publ. No. 3-A. 548 p.

McKinnell, R. G. 1962. Intraspecific nuclear transplantation in frogs. J. Hered. 53:199-207.

McKinnell, R. G. 1964. Expression of the kandiyohi gene in triploid frogs produced by nuclear transplantation. Genetics 49:895-903.

McKinnell, R. G. 1965. Incidence and histology of renal tumors of leopard frogs from the North Central states. Ann. N.Y. Acad. Sci. 126:85-98.

McKinnell, R. G. 1973. The Lucké frog kidney tumor and its herpesvirus. Am. Zool. 13:97-114.

McKinnell, R. G., and D. C. Dapkus. 1973. The distribution of burnsi and kandiyohi frogs in Minnesota and contiguous states. Am. Zool. 13:81-84.

McKinnell, R. G., and V. L. Ellis. 1972. Epidemiology of the frog renal tumour and the significance of tumour nuclear transplantation studies to a viral aetiology of the tumour, p. 183-197. *In* P. M. Briggs, G. de-Thé, and L. N. Payne [ed.] Oncogenesis and herpesviruses. International Agency for Research on Cancer, Lyon.

McKinnell, R. G., M. F. Mims, and L. A. Reed. 1969. Laser ablation of maternal chromosomes in eggs of *Rana pipiens*. Z. Zellforsch. 93:30-35.

Miles, E. M. 1950. Red-leg in tree frogs caused by *Bacterium alkaligenes*. J. Gen. Microbiol. 4:434-436.

Mizell, M. [ed.]. 1969. Biology of amphibian tumors. Springer-Verlag, New York. 484 p.

Mizell, M., I. Toplin, and J. J. Isaacs. 1969. Tumor induction in the developing frog kidneys by a zonal centrifuge purified fraction of the frog herpes-type virus. Science 165:1134-1137.

Moore, J. A. 1942. An embryological and genetical study of *Rana burnsi* weed. Genetics 27:406-416.

Moore, J. A. 1955. Abnormal combinations of nuclear and cytoplasmic systems in frogs and toads. Adv. Genet. 7:139-182.

Nace, G. W. 1968. The amphibian facility at the University of Michigan. BioScience 18:767-775.

Nace, G. W. 1970. The use of amphibians in biomedical research, p. 103-124. *In* Charles C. Middleton (chairman) Animal models for biomedical research III. Proceedings of a Symposium. National Academy of Sciences, Washington, D.C.

Nace, G. W., and C. M. Richards. 1969. Development of biologically defined strains of amphibians, p. 409-418. *In* M. Mizell [ed.] Biology of amphibian tumors. Springer-Verlag, New York.

Nace, G. W., and C. M. Richards. 1972a. Living frogs. 1. Adults. Carolina Tips 35:37-39.

Nace, G. W., and C. M. Richards. 1972b. Living frogs. 2. Care. Carolina Tips 35: 41-43.

Nace, G. W., and C. M. Richards. 1972c. Living frogs. 3. Tadpoles. Carolina Tips 35:45-47.

Nace, G. W., C. M. Richards, and T. Kawamura, 1965. Development of an amphibian facility. Anim. Care Panel 17:80.

Nace, G. W., C. M. Richards, and H. Sambuichi. 1966. Establishment of an amphibian facility. Am. Zool. 6(4):547.

Nace, G. W., C. M. Richards, and J. H. Asher, Jr. 1970. Parthenogenesis and genetic variability. I. Linkage and inbreeding estimations in the frog, *Rana pipiens*. Genetics 66:349-368.

Nace, G. W., J. K. Waage, and C. M. Richards. 1971. Sources of amphibians for research. BioScience 21:768-773.

Nace, G. W., C. M. Richards, and G. M. Hazen. 1973. Information control in the amphibian facility: The use of *R. pipiens* disruptive patterning for individual identification and genetic studies. Am. Zool. 13:115-137.

Napier, E. A. 1968. Transport of trigliceride across the rat mucosa. Fed. Proc. 27:635.

Nieuwkoop, P. D., and J. Faber [ed.]. 1956. Normal table of *Xenopus laevis* (Daudin); A systematical and chronological survey of the development from the fertilized egg till the end of metamorphosis. North-Holland Publ., Amsterdam. 359 p.

Orton, G. L. 1952. Key to the genera of tadpoles in the United States and Canada. Am. Midl. Nat. 47:382-412.

Pace, A. 1972. Systematic and biological studies of the leopard frogs (*Rana pipiens* complex) of the United States. Ph.D. thesis, University of Michigan, Ann Arbor.

Papermaster, D. S., and E. Gralla. 1973. Frog health. Science 180:10.

Pickering, Q. H., and W. N. Vigor. 1965. The acute toxicity of zinc to eggs and fry of the fathead minnow. Prog. Fish-Cult. 27:153-157.

Platz, J. E., and A. L. Platz. 1973. *Rana pipiens* complex: Hemoglobin phenotypes of sympatric and allopatric populations in Arizona. Science 179:1334-1336.

Poiley, S. M. 1972. Growth tables for 66 strains and stocks of laboratory animals. Lab. Anim. Sci. 22:759-779.

Pollister, A. W., and J. A. Moore. 1937. Tables for the normal development of *Rana sylvatica*. Anat. Rec. 68:489-496.

Porter, K. R. 1939. Androgenetic development of the egg of *Rana pipiens*. Biol. Bull. 77:233-257.

Porter, K. R. 1972. Herpetology. W. B. Saunders Co., Philadelphia. 524 p.

Priddy, J. M., and D. D. Culley. 1971. The frog culture industry, past and present. Proc. 25th Annu. Conf. S.E. Assoc. Game Fish Comm., Charleston, S.C.:597-601.

Rafferty, Jr., K. A. 1962. Age and environmental temperature as factors influencing development of kidney tumors in uninoculated frogs. J. Natl. Cancer Inst. 29:253-265.

Rafferty, Jr., K. A. 1965. The cultivation of inclusion associated viruses from Lucké tumor frogs. Ann. N.Y. Acad. Sci. 126:3-26.

Reichenbach-Klinke, H., and E. Elkan. 1965. The principal diseases of lower vertebrates. Academic Press, New York. 600 p.

Reid, G. K. 1961. Ecology of inland waters and estuaries. Van Nostrand Reinhold, New York. 375 p.

Richards, C. M. 1958. The inhibition of growth in crowded *Rana pipiens* tadpoles. Physiol. Zool. 31:138-151.

Richards, C. M. 1962. The control of tadpole growth by algae-like cells. Physiol. Zool. 35:285-296.
Richards, C. M., and G. W. Nace. In press. Sex in *Rana pipiens*. Test of the XX-XY hypothesis based on unusual male frequencies among gynogenetic and certain biparental progeny and on hormonal sex reversal. Submitted to J. Environ. Zool.
Richards, C. M., D. T. Tartof, and G. W. Nace. 1969. A melanoid variant in *Rana pipiens*. Copeia (4):850-852.
Richardson, L. R. 1937. Observations on trichodinid infection [Cyclochaetosis of *Salvelinus fontinalis* (Mitchill)]. Trans. Am. Fish. Soc. (Washington, D.C.) 67:228-231.
Rose, S. M. 1946. Disease control in frogs. Science 104:330.
Rose, S. M., and F. C. Rose. 1965. The control of growth and reproduction in freshwater organisms by specific products. Mitt. Int. Ver. Limnol. 13:21-35.
Rowlands, Jr., D. T. 1969. General mechanisms and principles of immunity in cold-blooded vertebrates, p. 231-248. *In* G. J. Jackson, R. Herman, and I. Singer [ed.] Immunity to parasitic animals, Vol. 1. Appleton-Century-Crofts, New York.
Rugh, R. 1965. Experimental embryology. Techniques and procedures, 3rd ed. Burgess, Minneapolis. 501 p.
Russel, F. H. 1898. An epidemic, septicemic disease among frogs due to the *Bacillus hydrophilus fuscus*. J. Am. Med. Assoc. 20:1442-1449.
Sambuichi, H. 1959. Production of polyploids by means of transplantation of nuclei in frog eggs. J. Sci. Hiroshima Univ., Ser. B., Div. 1 18:39-43.
Sarkar, H. B. D., and M. Appaswamy Rao. 1971. Effect of thyroidectomy and administration of thyroxine on ovulation and spawning *in vivo, in vitro* and in transplantation in the skipper frog, *Rana cyanophlyctis* (Schn.). Gen. Comp. Endocrinol. 16:594-598.
Savage, R. M. 1961. The ecology and life history of the common frog. Sir Isaac Pitman & Sons, Ltd., London. 221 p.
Schmidt, R. S., and W. R. Hudson. 1969. Maintenance of adult anurans. Lab. Anim. Sci. 19:617-620.
Schuetz, A. W. 1967. Action of hormones on germinal vesicle breakdown in frog (*Rana pipiens*) oocytes. J. Exp. Zool. 166:347-354.
Schwabe, C. W. 1969. Veterinary medicine and human health, 2nd ed. Williams & Wilkins, Baltimore. 713 p.
Shaver, J. R. 1953. Studies on the initiation of cleavage in the frog egg. J. Exp. Zool. 122:169-192.
Shumway, W. 1940. Stages in the normal development of *Rana pipiens*. I. External form. Anat. Rec. 78:139-144.
Signoret, J., R. Briggs, and R. R. Humphrey. 1962. Nuclear transplantation in the axolotl. Develop. Biol. 4:134-164.
Smith, H. M. 1969. The Mexican axolotl: Some misconceptions and problems. BioScience 19:593-597.
Smith-Gill, S. J., C. M. Richards, and G. W. Nace. 1972. Genetic and metabolic bases of two "albino" phenotypes in the leopard frog, *Rana pipiens*. J. Exp. Zool. 180:157-167.
Society for the study of Amphibians and Reptiles. 1971 *et seq*. Catalogue of American amphibians and reptiles. American Society of Ichthyologists and Herpetologists, American Museum of Natural History, New York. (loose-leaf)

Spotte, S. H. 1970. Fish and invertebrate culture—Water management in closed systems. Wiley-Interscience, New York. 145 p.

Staats, J. 1972. Standardized nomenclature for inbred strains of mice: Fifth listing. Cancer Res. 32:1609–1646.

Stearns, J. E. 1973. How it all began. Published by author, P.O. Box 514, Palestine, Texas 75801. 60 p.

Stebbins, R. C. 1966. A field guide to western amphibians and reptiles. Houghton-Mifflin, Boston. 279 p.

Stober, Q. J., and W. R. Payne, Jr. 1966. A method for preparation of pesticide-free fish food from commercial fish food pellets. Trans. Am. Fish. Soc. 95: 212–213.

Subtelny, S. 1958. The development of haploid and homozygous diploid frog embryos obtained from transplantations of haploid nuclei. J. Exp. Zool. 139:263–305.

Subtelny, S., and C. Bradt. 1963. Cytological observations on the early developmental stages of activated *Rana pipiens* eggs receiving a transplanted blastula nucleus. J. Morphol. 112:45–60.

Taylor, A. C., and J. J. Kollros. 1946. Stages in the normal development of *Rana pipiens* larvae. Anat. Rec. 94:2–23.

Tweedell, K. S. 1967. Induced oncogenesis in developing frog kidney cells. Cancer Res. 27:2042–2052.

Tweedell, K. S. 1972. Experimental alteration of the oncogenicity of frog tumor cell-viral fractions. Proc. Soc. Exp. Biol. Med. 140:1246–1251.

Tweedell, K. S., and A. Granoff. 1968. Viruses and renal carcinoma of *Rana pipiens.* V. Effect of frog virus 3 on developing frog embryos and larvae. J. Natl. Cancer Inst. 40:407–410.

Van der Hoeden, J. [ed.]. 1964. Zoonoses. Elsevier Publishing Co., Amsterdam. 774 p.

Van der Steen, A. B. M., B. J. Cohen, D. H. Ringler, G. D. Abrams, and C. M. Richards. 1972. Cutaneous neoplasms in the leopard frog (*Rana pipiens*). Lab. Anim. Sci. 22:216–222.

van der Waaij, D., T. M. Speltie, and J. M. Vossen. 1972. Biotyping of Enterobacteriaceae as a test for the evaluation of isolation systems. J. Hyg. Camb. 70:639–650.

van der Waaij, D., B. J. Cohen, and G. W. Nace. In press. Colonization patterns of aerobic gram negative bacteria in the cloaca of *Rana pipiens.* J. Lab. Anim. Sci. 24.

Volpe, E. P. 1955. A taxo-genetic analysis of the status of *Rana kandiyohi* weed. Syst. Zool. 4:75–82.

Volpe, E. P. [ed.]. 1971. Biology of immunity in amphibians. Am. Zool. 11:167–237.

Volpe, E. P., and R. G. McKinnell. 1966. Successful tissue transplantation in frogs produced by nuclear transfer. J. Hered. 57:167–174.

Wagner, E. K., B. Roizman, T. Savage, P. G. Spear, M. Mizell, F. E. Durr, and D. Sypowicz. 1970. Characteristics of the DNA of herpesviruses associated with Lucké adenocarcinoma of the frog and Burkitt lymphoma of man. J. Virol. 42:257–261.

Walton, A. C. 1964. The parasites of amphibia. Wildlife disease, #39 and #40. (Microcards available through Wildlife Diseases Association, 333 N. Mich. Ave., Chicago, Ill.).

Walton, A. C. 1966. Supplemental catalog of the parasites of amphibia. Wildlife disease, #48.

Walton, A. C. 1967. Supplemental catalog of the parasites of amphibia. Wildlife disease, #50.

Weiz, P. B. 1945a. The development and morphology of the larvae of the South African clawed toad, *Xenopus laevis*. I. The third-form tadpole. J. Morphol. 77:163-191.

Weiz, P. B. 1945b. The development and morphology of the larvae of the South African clawed toad, *Xenopus laevis*. II. The hatching and the first-and-second form tadpoles. J. Morphol. 77:193-217.

Whipple, H. E. [ed.]. 1965. Viral diseases of poikilothermic vertebrates. Ann. N.Y. Acad. Sci. 126:1-680.

Witschi, E. 1930. Studies on sex differentiation and sex determination in amphibians. J. Exp. Zool. 56:149-165.

Wolf, D. P., and J. L. Hedrick. 1971. A molecular approach to fertilization. II. Viability and artificial fertilization of *Xenopus laevis* gametes. Devel. Biol. 25:348-359.

Wolf, K., G. L. Bullock, C. E. Dunbar, and M. C. Quimby. 1968. Tadpole edema virus: A viscerotropic pathogen for anuran amphibians. J. Infect. Dis. 118:253-262.

Wolf, K., G. L. Bullock, C. E. Dunbar, and M.C. Quimby. 1969. Tadpole edema virus: Pathogenesis and growth studies and additional sites of virus infected bullfrog tadpoles, p. 326-336. *In* M. Mizell [ed.] The biology of amphibian tumors. Springer-Verlag, New York.

Woolley, H. P. 1973. Subcutaneous acrylic polymer injections as a marking technique for amphibians. Copeia (2):340-341.

Wright, A. H. 1920. Frogs: Their natural history and utilization. Bureau of Fisheries Document No. 888. Government Printing Office, Washington, D.C.

Wright, A. H., and A. A. Wright. 1949. Handbook of frogs and toads. Comstock, Ithaca, New York. 640 p.

Wright, P. A., and A. R. Flathers. 1961. Facility of pituitary induced ovulation by progesterone in early fall. Proc. Soc. Exp. Biol. Med. 106:346-347.

Zambernard, J., and A. E. Vatter. 1966. The fine structural cyto-chemistry of virus particles found in renal tumors of leopard frogs. I. An enzymatic study of the viral nucleoid. Virology 28:318-324.

To assist ILAR in determining the usefulness of this guide, please complete and return this questionnaire to the Institute of Laboratory Animal Resources, National Academy of Sciences, 2101 Constitution Avenue, Washington, D.C. 20418.

QUESTIONNAIRE

Name

Institution

Department

1. Specific area of scientific interests?

2. What portions of the guide did you find particularly useful?

    a. Which sections did you find difficult to use?

    b. Were any sections contrary to your experience or misleading?

    c. Were any sections too vague or overly explicit?

3. Suggestions for the improvement of future issues:

    a. Format:

    b. Contents:

    c. Other:

4. Purpose for which guide was obtained:
    General reference _____
    Specific requirement for information _____
    Other (specify) _____

Amphibians